The surface of the earth:
an introduction to geotechnical science

Peter J. Williams

Professor of Geography and
Director, Geotechnical Science Laboratories,
Carleton University

Longman
London and New York

Longman Group Limited
Longman House
Burnt Mill, Harlow, Essex, UK

*Published in the United States of America
by Longman Inc., New York*

First published 1982

British Library Cataloguing in Publication Data

Williams, Peter J.
 The surface of the earth.
 1. Earth sciences
 I. Title
 551.1   QE26.2

 !SBN 0-582-30043-6

Library of Congress Cataloging in Publication Data

Williams, Peter J.
 The surface of the earth.

 Bibliography: p.
 Includes index.
 1. Earth sciences.   2. Engineering geology.
 I. Title.
 QE33.W54   551   81-3683
 ISBN 0-582-30043-6   AACR2

Printed in Great Britain by Pitman Press Ltd., Bath

Contents

Preface

The explanation of what we observe in the natural environment immediately around us, as well as globally, lies in the behaviour of the individual gaseous, liquid or solid components of the earth's surface. Axiomatic though this may seem, most books in physical geography, or other fields pertaining to the environment, concentrate instead upon the composite characteristics of localities or regions. Yet landscapes, terrain or climate represent the interaction of many different phenomena and materials, and this is also true of quite small features – a small area of ground surface or a particular hillslope, for example. This book is concerned with understanding the natural environment through a detailed examination of what may be called, by analogy with 'microclimate', the 'microenvironment'. This involves consideration of specific material properties and processes.

It is hoped that the book will be useful in respect to a wide range of interests. Architects and engineers normally have a basic understanding of properties of earth materials, but a limited knowledge of how these relate to the characteristics of a locality. The importance of the sequence of events through geological time, in producing existing landforms, is not always realized by those unfamiliar with the earth sciences. For these reasons, the first part of the book includes an elementary review of fundamental principles of geomorphology and surface geology, which may also be useful to beginning students in the earth sciences. The second part of the book considers mechanical properties of earth materials, and properties relative to exchanges of heat and moisture including those of the ground surface. The topics considered are not usually brought together in one book, even though they cannot properly be considered in isolation. Design of energy-efficient buildings, concerns with food production and desertification, and geotechnical development in cold regions are all examples of the diverse technological issues which require a balanced knowledge of thermal and other energy-associated properties, together with a knowledge of the mechanics of earth materials. The quantitative information included will, it is hoped, make the book of value as a reference work. The third part of the book considers various environmental situations, slopes, the ground immediately below the surface, very cold and very hot conditions and demonstrates how scientific interpretations of the environment follow from a precise knowledge of material properties and processes.

University students in geography and geology and in the biological sciences, indeed in the environmental sciences as a whole, often do not have a strong background in the basic sciences and mathematics. Accordingly care has been taken to explain carefully the physical, chemical and thermodynamic concepts used in the text. With a few exceptions which may be passed over if desired, only simple equations are used. The book is intended for students at the undergraduate level, while elements of the book are useful for higher level, more specialized studies. The subject matter and arrangement of the book has been much influenced by my experiences during the last ten years in teaching students, from the second year to post-graduate levels, in the Physical Geography and Geotechnical Science program at Carleton University.

In addition to the many students who have worked with chapters of the book in courses that I have taught, providing many helpful comments, I am indebted to many people for assistance through the last six years. The book was initially drafted during a sabbatical leave spent at the Scott Polar Research Institute,

Cambridge, and I am especially indebted to Dr David Drewry for his comments on the early drafts, and to Dr Gordon Robin for his hospitality as Director of the Institute. I have had many enjoyable and productive discussions with Dr Donald Davidson of the University of Strathclyde, who shares my interest in introducing physical science into matters geographical and geotechnical. My colleagues in the Geotechnical Science Program at Carleton University, especially Dr M. W. Smith, and Dr B-E. Ryden (a visitor from Upsala University), have kindly read chapters and provided stimulus in discussions for many years. My father J. G. Williams has provided many comments as an interested layman, which have led to improvements in the clarity of the text, and he has also prepared the index. My successive, able and considerable typists Jane Whiting, Anne Buie and Janet Wilson, have been paid largely through a small grant from the National Research Council of Canada – one of the first for such a purpose. I hope the book will help justify this policy. Activities both in university teaching and geotechnical consulting, have prolonged the writing of this book, probably with advantage. My contacts with many professional colleagues too numerous to name, have been an important influence in defining the content of the book.

Peter J. Williams
Ottawa
Canada
April 1981

Symbols

A	area
a	thermal diffusivity
C	volumetric heat capacity; cohesion
c_s	mass heat capacity; soil mineral
c_v	coefficient of consolidation
c_w	mass heat capacity; water
D	soil water diffusivity
e	void ratio of soils
F	factor of safety; frost index
G	Gibbs free energy; heat flux in or out of ground
g	acceleration due to gravity (= 9.8 m s^{-2}); gram
H	enthalpy; sensible heat flux
h or H	vertical distance above or below water table; sensible heat flux
i	gradient of potential; angle of slope
$I\downarrow$	radiation, inwards
$I\uparrow$	radiation, outwards
k	hydraulic conductivity
K	degrees Kelvin
L	latent heat of freezing per unit volume soil; length
LE	heat flux associated with evaporation or condensation
ℓ	distance between points of interest; latent heat of fusion; length; litre
n	porosity
P	pressure
P_c	preconsolidation pressure
P_o	saturation vapour pressure for pure, free water
Q	incoming solar radiation; total heat flow
Q_{vap}	vapour flux
q	flow; incoming diffuse radiation
q	mean specific humidity
R	net radiative flux; universal gas constant
r	run-off; reflected radiation
SI	Systeme Internationale
S_r	degree of saturation
S	strength
T	temperature; absolute temperature; time
t	time period; time; temperature
U	pore water pressure; wind speed
u	pore water pressure
u	average wind velocity
V	volume
v	specific volume

Z	depth	σ	normal stress; Stefan-Boltzman constant
β	Bowens ratio; angle of slope	$\acute{\sigma}$	effective normal stress
γ	unit weight	τ	shear stress
γ_d	dry unit weight	τ_s	shear stress when shear occurs
γ_w	unit weight of water	ϕ	angle, the tangent of which is the
ε	emissivity		coefficient of proportionality of shear
θ	moisture content; contact angle		strength and normal stress – 'angle of
λ	thermal conductivity coefficient		internal friction'
ρ	density; soil bulk density	ψ	potential
ρ_d	dry bulk density	μ	micron
ρ_w	density of water		

Units

Units in this book follow the SI System (Systeme Internationale). Conversions between the SI System and other systems are given in, for example, Kaye, G. W. C. and Laby, T. H., 1973, *Tables of Physical and Chemical Constants*, Longman, 386 pp. A convenient guide to SI usage in geotechnical engineering is given in the *Proceedings, 9th International Conference on Soil Mechanics and Foundation Engineering*, 1977, Vol. 3, pp. 153–70 where internationally accepted symbols for use in geotechnical studies are also listed.

Acknowledgements

We are grateful to the following for permission to reproduce copyright material:

American Elsevier Publishing Co. Inc. for our fig. 8.3 from fig. 5.1, our fig. 8.4 from fig. 5.4 and our fig. 8.5 from fig. 5.6 (Monteith 1973); American Meteorological Society for our fig. 8.1 from fig. 1 *Meteorological Monographs* Vol. 6, No. 28, 1965; American Society of Agricultural Engineers for our fig. 7.2 modified from fig. 1 (Mather 1959); Artemis Press Ltd. for our fig. 1.4 from fig. 12.5 (Goldring 1971); Associated Book Publishers Ltd. for our fig. 3.3 from fig. 2 & 5 (Lamb 1966); B. T. Batsford Ltd. for our fig. 12.1 from fig. 2.7 and our fig. 12.6 from fig. 4.5 (Cooke & Warren 1973); Professor G. H. Bolt for our fig. 6.8 a & b from fig. 14a & b (Bolt 1970); Cambridge University Press for our fig. 9.3 from fig. 7.11 (Carson & Kirkby 1972); Dr Jen-Hu Chang for our figs. 8.6 & 8.7 from figs. 59 & 61 (Chang 1968); Elsevier Scientific Publishing Co. for our fig. 10.7 & plate 10.1 from figs. 9 & 10 (Crawford 1968); Elsevier Scientific Publishing Co. and John Wiley & Sons Inc. for our figs. 10.4 & 10.5 from figs. 1, 3, 5 & 6 (Holmes & Colville 1970) and figs. 10.2 & 10.3 (Baver, Gardner & Gardner 1972); Geological Survey of Sweden for our fig. 10.2b from fig. 79 (Beskow 1935); Harvard University Press for our fig. 3.1 from fig. 117 (Geiger 1965); Hemisphere Publishing Corporation for our fig. 11.12 from fig. 81 (Tsytovich 1973); Journal of Soil Science Society of America for our fig. 10.6 from figs. 8 & 9 (Staple 1969); Keter Publishing House Jerusalem Ltd. for our fig. 8.2 from fig. 127 (Petrov 1976) English translation by John Wiley & Sons Inc; Longman Group Ltd. for our fig. 8.8 from fig. 2.3 (Williams 1979), our fig. 11.1 (modified after Brown 1970) from fig. 1.2, our fig. 11.3 (modified after Brown 1970) from fig. 1.5 and our fig. 11.10 from fig. 2.2 (Williams 1979), our fig. 12.4 from fig. 2.1 (Morgan 1979); Macmillan Publishing Co. Inc. for our fig. 6.6 from fig. 7.6 (Brady 1974) Copyright © 1974 by Macmillan Publishing Co. Inc; Morgan Grampian Ltd. for our plate

6.1 from fig. 10 (Williams 1967); National Academy of Sciences for our fig. 11.2 from fig. 8 (Brown 1973) and fig. 6 (Lachenbruch 1963); National Research Council of Canada for our fig. 11.4 from fig. 20 (Mackay 1973) and our fig. 9.6 from fig. 3 (Mitchell & Eden 1972); Norwegian Geotechnical Institute for our fig. 9.4 from fig. 2 and our fig. 9.5 from fig. 6 (Bjerrum, Löken, Heiberg & Foster 1969); Oliver and Boyd and Plenum Publishing Corporation for our fig. 12.2 from fig. 2 (Strakhov 1967); Oxford University Press for our fig. 4.9 from fig. 2.13 (Stratham 1977); Royal Geographical Society for our fig. 11.7 from fig. 1 (Williams 1957); Royal Microscopical Society for our plate 2.1 (Gillott 1980); R. Sätersdal for our fig. 5.3 from fig. 1 (Johansen 1973); Society Mexicana de Mecanica de Suelas for our fig. 9.2 from figs. 1 & 2 (Skempton & Hutchinson 1969); Mr W. J. Staple and the Minister of Supply and Services, Canada for our fig. 6.11 from fig. 2 (Staple 1967); Swedish Natural Science Research Council for our fig. 11.6 from fig. 4 (Beskow 1935); Transport and Road Research Laboratory for our fig. 7.3 modified from fig. 21 (Croney, Coleman & Bridge 1952); University of Chicago Press for our fig. 3.2 from figs. 30 & 31 (Sellers 1965), our fig. 7.1 from fig. 2.2 (Byers 1965); University of Guelph for our plate 11.1, our fig. 10.1 from fig. 1 and our fig. 10.3 from fig. 2 (Williams & Nickling 1972); Van Nostrand Reinhold Co. for our fig. 2.1 from fig. 1.7 (Craig 1974); John Wiley & Sons Inc. for our fig. 12.3 from fig. 3.5 (Sanchez 1976), our fig. 6.10 from part of fig. 19.5 (Lambe & Whitman 1979), our fig. 2.4 from fig. 2.5 (Curtis 1976), our fig. 1.1 from fig. 2.1, our fig. 1.2, from fig. 6.1 and our fig. 1.3 from fig. 6.7 (Wyllie 1976); Mr G. P. Williams for our table 5.3 from table II (Williams 1970); Mr N. P. Woodruff and Academic Press Inc. for our fig. 12.5 from fig. 1 (Chepil & Woodruff 1963).

PART I

The earth's surface as a study in dynamics

1

General nature of the earth's surface

1.1 Introduction

This book is concerned with a thin layer at the surface of the earth. The layer has a thickness less than 0.01 per cent of the earth's radius, and extends only a few metres or tens of metres above and below the ground surface. Nevertheless, it is this layer which constitutes Man's environment in the ecological sense; it provides both the resources and conditions suitable for life as well as the restraints and boundaries which Man by his activities seeks to overcome. Only exceptionally does Man move upwards out of this thin layer, and then he can usually only do so by effectively moving his local environment with him, as in high-flying aircraft. Equally sharp changes in the natural environment occur on moving downwards, for at a depth of 1 kilometre the temperature of the ground has risen by about 25 °C, and at 10 kilometres by 250 °C (Sass 1971).

The approach in this book involves the application of fundamental scientific principles or laws and these of course are equally applicable to the centre of the earth as to the upper atmosphere. But away from the surface regions of the earth, pressures, temperatures, and the nature of materials are substantially different. The interior regions are considered therefore only briefly and in so far as this helps to providing a background for our analysis. Those properties of earth materials which control the natural processes that man directly experiences and to which, frequently, he must react are, together with the processes themselves, the subject matter of this book.

The history of natural science is that of man's inquiry into his physical surroundings. Over the last few centuries much of natural science has involved those natural phenomena, or properties of matter, which can be investigated experimentally, using samples of a larger whole. Yet remarkably, even in this century, much of the earth's natural surface has been investigated in the field, or from the study, without recourse to the laboratory. More precisely, while the surface forms were measured and recorded, and the earth materials were subjected to gross description and classification, scant regard was paid to the large range of mechanical and thermodynamic properties possessed by those materials, let alone to the range of dynamic situations resulting from those properties, even within a small area of ground.

Early earth scientists, once freed of the philosophical and theological inhibitions of the Middle Ages (which persisted in some degree to the nineteenth century), directed their attention to whichever particular natural phenomenon at hand fascinated them. Curious patterns and forms in the rocks were ultimately recognised as fossils. Their marine origin being no longer in dispute, a further period elapsed before their presence at high elevations was understood to be the result of effects still active today. The rejection of abnormal and catastrophic events, such as the biblical flood, in favour of processes that can be observed led to the formulation of Hutton's 'Principle of Uniformitarianism' – that the present is the key to the past. Geological explanation must necessarily be limited if we have only an incomplete understanding of the land forming processes as they are at present observed. We must first educate ourselves about the nature and behaviour of the various materials of which the earth's surface is composed. This will in turn require an understanding of geological history; thus the circle is closed.

Particularly during the nineteenth century, a new type of scientist appeared. His studies were directed towards the soils and rocks, but his interest was motivated by practical aims, of agriculture (Hutton himself operated an experimental farm), of civil engineering, and of mining and quarrying. It took hundreds of years, for science to develop from an exclusively academic pursuit into a largely utilitarian branch of knowledge. In doing so, the development of these practical aims has not only influenced the nature of the knowledge being gained – it has also influenced, sometimes unfortunately, the quality of that knowledge.

Today, soils and rocks enter into the activities of many occupational groups. Foundation engineering, the construction of big buildings, bridges, dams, requires an intimate knowledge of the earth materials, in particular their so-called geotechnical properties. Because the agriculturalist, and the engineer for example are to some extent concerned with the same materials we would expect there to be a substantial body of common knowledge. But, an examination of the terminologies of the published literature, of the organization of universities and research institutes which relate to 'agricultural' soils, and of those which relate to 'engineering' soils, would suggest to those who did not know otherwise that these two groups of materials were quite separate, both in form and function.

Many other examples could be given of how the 'tunnel vision' of scientists is perpetuated by traditional packaging into 'subjects', whether by the effects of history or of practical ends. This is unfortunate because, whatever we study in the so-called 'derivative' or 'secondary' sciences – biology, geology, pedology, hydrology – we must always operate within the established principles and laws of the primary sciences of mathematics, physics and chemistry. We can only benefit by realizing that others may have new and different insights into the phenomena we study in our own 'field'. The integral nature of scientific studies must not be overlooked even though in practical applications such as agronomy, forestry, civil engineering, there may be different emphases. These, and other occupations, all need to know about the materials of the earth's surface region.

This book may be regarded by some as a text in geomorphology, but it could also be considered to be in geotechnical science, that is, describing scientific studies which are ultimately prompted by man's interaction with his surroundings through engineering and other activities. The workings of the earth surface are considered through an analysis of composite parts. The approach is on a microscale; we consider the properties and behaviour of small and well-defined elements of the whole.

Ultimately, the aim is to permit an understanding of why for example, a landslide occurred, why the ground is frozen throughout the summer at one small location but not at neighbouring sites; or why water moves spontaneously from an apparently less wet sand into a wetter clay.

To a great extent we are concerned with findings originally obtained from samples in a laboratory. These findings are then applied to the interpretation of the various processes that define the physical characteristics of the earth's surface region at a particular place. The latter qualification is important. A single description and analysis of this type cannot be applied simultaneously to a large area. This does not mean that a basic understanding of the grosser characteristics and behaviour of the earth's surface region is not important. Such understanding is often essential to the interpretation of the local event or feature, and this and the following chapter provide an elementary review of the global framework.

1.2 Some basic energy relations of the earth's surface

Volcanic lava from the interior of the earth is molten, but the naturally occurring temperatures at the surface of the earth are lower than the melting points of all the common minerals there. The earth as a whole is cooling slowly, because there is a flow of heat from the depths towards the surface which although varying considerably from place to place, has values (Lee and Clark 1966) around 0.05 W m^{-2} (that is, 0.05 watts through a square metre). The sun is radiating energy towards the earth at a power of 10^{17} W of which about 10^{16} W reaches the solid surface (Von Arx 1974). A very rough calculation, based on the area in square metres of the land surface of the earth (approximately 10^{14} m^2) suggests that more solar energy is arriving (at a power of about 100 W m^{-2}) than heat is being lost from the earth (about 0.05 W m^{-2}). Indeed

the arrival of energy from above and below might lead us to assume that there must be an accumulation of energy in the earth's surface region. If so, it can hardly be solely in the form of heat for the temperature of the ground would be rising rapidly.

Before examining the possibilities more closely the terms energy and power must be considered. *Energy* is defined as the capability to do work. *Work* is said to be done when a force acts against a resistance to produce motion. Force is defined simply as 'that which changes the state of rest or motion'. The amount of work is the product of the force and the distance moved. We may think of a block sliding down a sloping surface. It is gravity which causes it to slide while the magnitude of the force actually effective in producing the motion depends on the slope as well as the weight of the body. The evaluation of the force will be discussed later. Force is measured in newtons, (N), and if the force to produce sliding is X newtons, on moving the block a distance L metres downslope, work XL newton metres is done. Newton metres are units of work and are represented by the symbols N m. They are also known as joules (symbol: J).

If the block is lifted from the surface after sliding distance L, and replaced at its initial position on the slope, we restore its ability to slide over L. Its capability of performing the work is restored. This increment of energy is called *potential energy*, and potential energy is that associated with position. In this case we are concerned with gravitational potential energy which results from position in the earth's gravitational field. This increment of energy will be utilized when the block again slides through L.

If the distance L is traversed in a certain time t s (i.e. seconds), the work is clearly being done at a rate of $\dfrac{XL}{t}$ N m s^{-1}, or $\dfrac{XL}{t}$ J s^{-1}. If XL has a different value work is being done at a different rate. The rate at which work is done is called the *power*. The power is therefore the rate of energy dissipation. It is equally the rate at which energy is being supplied for dissipation. Thus when we speak of the power of the sun, we are speaking of that part of the energy dissipated by the sun, that is reaching the earth where it is again dissipated.

If we again consider the sliding block, we would expect, if we had an accurate enough thermometer, to observe a temperature rise at the base of the block due to friction. Heat is a form of energy, and part of the energy being dissipated is going to heating. The fact that energy may be dissipated as work or as heat means that work and heat have a fundamental equality. When we say that energy is dissipated we do not mean that it disappears. On the contrary a fundamental principle is that of the conservation (or indestructibility) of energy. Dissipation refers to energy in transit and this is revealed as either heat or work. In the course of the transit, the one may be changed into the other (even though a fundamental law tells us that a simple 100 per cent conversion of heat to work is not possible). Work was not originally, and is not always, measured in the same units as heat. But with the acceptance in recent years of the 'Systeme Internationale' (SI) system of units the relationship is made clearer. The unit of energy is the joule, which is the unit of heat as well as of work, and 1 J s^{-1} is called a watt.

The transfer of 4.18 J in the form of heat, to a gram of water causes its temperature to rise by *1* °C, and likewise, when a gram of water falls in temperature by 1 °C, 4.18 joules have in some manner been lost from it. Reflection over this suggests that temperature is analogous in this respect to potential energy. In fact when the temperature is changed there is a change of *internal energy*.

It is not possible, in this brief introduction, to consider these concepts in detail. The reader is referred to Davidson 1978, and to basic texts in physics and thermodynamics of which examples are given in the bibliography. However, an elementary grasp of the relationships between energy, power, work and heat are sufficient for further considerations of the effects of solar radiation, and also of the other energy sources which are involved in the processes operating at the earth's surface. In addition, the more detailed considerations of the properties of earth materials in Part II will make the understanding more precise.

Of the energy arriving from the sun, some is immediately returned to space by reflection in the atmosphere from clouds and other particles with reflecting surfaces, and some is scattered (or deflected) by the gaseous constitutents of the air. Absorption (that is conversion to heat) by constituents of the atmosphere also takes place. Reflection of a further part of the incoming solar radiation occurs at the surface of the earth. The amount reflected depends on the reflectivity or *albedo* of the surface. Snow reflects much radiation, while dark surfaces, bare soil for example, reflect less, having a lower albedo. The albedo of the earth's surface is usually determined by the vegetation cover, since this commonly shields much of the soil. The radiation, reaching the earth's surface from the sun, which is not reflected is energy that is dissipated in various ways. It may cause a warming of the ground, the plant leaves, or the air and is then referred to as 'sensible heat': the heating effect can be measured as a temperature change. During night time temperatures fall because there is radiation outwards. This does not constitute reflection, since the energy has been converted to internal energy, and then re-emitted. Furthermore the emitted radiation includes not only energy received the previous day, but also perhaps the previous summer, or much earlier. Some of the energy being lost must be that associated with the cooling of the interior of the earth. Of course the energy cannot be 'tagged' – we cannot usually physically recognise energy by some characteristic which reveals that it has its origin, for example, in the centre of the earth. Nevertheless when we seek to explain certain observed temperature changes, or certain work done, it is necessary to compute the magnitudes of *all* the energy flows to or from the system.

This kind of analysis is referred to as the study of energy balances. An energy balance is simply an equation, in which as in a bank account, one identifies incomings, outgoings, and capital, the latter representing for example, storage of heat energy or potential energy. Once money is paid into the bank, it is no longer possible to identify those particular bank notes but their effect is clearly recognisable.

In addition to reflection, and sensible heat, some of the incoming energy is utilised in evaporation. To convert water to water vapour involves the addition of roughly 2400 10^3 J kg^{-1} (joules per kilogram). Contrary to common belief, evaporation commonly does not require a temperature rise. As long as evaporation is freely occurring there is a cooling effect on the surroundings as heat is transferred to the evaporating water. The reason water is warmed to boil it, is that at 100 °C and atmospheric pressure, water evaporates freely, immediately, and not merely at its surface.

Evaporation causes a radical change in the density of the water substance as it expands greatly on being converted to vapour. Movement is greatly increased in the vapour phase. Increasing temperature of course causes expansion, which gives rise to air movements, to movements of water substance, and ultimately to rainfall. This latter involves a reversal of the process of evaporation, and the reappearance of liquid water by condensation involves the loss from the water of roughly 2400 10^3 J kg^{-1}.

Movement of mass whether solid, liquid or gas, involves forces acting through distance, and thus constitutes work. In engineering, a machine that converts heat to work is called a heat engine. The earth's surface region involves natural heat engines of many different forms.

Thus we begin to see that the solar energy, constantly arriving at the surface of the earth, is dissipated in innumerable and complex ways. The same can be said of the much smaller quantity of energy arriving from the earth's centre in the form of a heat flow. Fortunately for mankind there is not a long-term accumulation of internal energy with a concomitant temperature rise at the earth's surface.

1.3 The forming of the land

When rain falls it frequently causes erosion, by dislodging and transporting soil particles. It is the potential energy of raindrops which is expanded in their fall, and, in part, in the motion which they give to soil particles.

The cumulative effects of rain and flowing water are in large measure responsible for the land forms around us. There are however two common misconceptions about the role of water in shaping land forms. The first is that it is the trickling of water over the ground surface which ultimately produces topographic form, and the second is that rivers by progressive erosion of their banks have cut their valleys. The ideas of course are to some extent contradictory. The former is erroneous in that water erosion over the surface of a slope is only one process, and often a minor one, involved in the movements of soil material downslope. The latter idea of the role of erosion by rivers is at best a misleading generalisation. A river erodes its bed. As the bed at the foot of a river bank is eroded, soil material above water level being no longer supported sufficiently tends to fall or slide into the river. Such movements spread upslope, as each slip leaves unsupported the next increment upslope. The *form* of the slope, the nature of the scars and other surface features left by the sliding or other movement of the material, cannot be said to be the characteristic product of river erosion. Landslides, mudflows and other displacements occur on the slope at points distant from the river without any direct and immediately preceding involvement of the river. Any other agency that removed the basal support of the slope, a bulldozer perhaps, would initiate similar effects. Within any slope there are forces of gravitational origin, which have the potential for causing movements of materials towards lower elevations. Such forces find expression in movements of many types. Counteracting them is the resistance or strength of materials comprising the slopes. The form of a slope depends on the properties of the materials of which the slope is composed, which in turn depend on the climate and the history of the slope.

Flowing water is an important, indeed fundamental, element in the overall process of denudation, which term is best regarded simply as the tendency towards lowering of the land surface. The effectiveness of the rivers and streams in removing material clearly depends on a variety of factors, such as the volume of flow, the gradient, the spatial arrangement of the drainage network and so on. Sometimes of course, erosion by a river leaves its direct imprint, as for example in flat-topped river terraces, cut successively by the river as it migrates from side to side and slowly down into the valley. These can usefully be regarded as 'erosional' landforms. But even in such cases, the form of the frontal edge or bluff of each terrace is dependent in large degree on the properties of the material composing the terrace, and the small, or large, slips and slumps down the bluff *after* this has been exposed from the water.

Both rivers and glaciers are also responsible, in a direct manner, for certain landforms known generally as depositional. Earth material which becomes incorporated in a river or glacier is carried along by the water or ice. In either case it may be carried without pause into the sea. Rather frequently however, conditions change downstream and the river or glacier can no longer carry its load which is then deposited. Often glaciers terminate before reaching the sea. The surface form of such deposited loads today, may result from events subsequent to deposition but not necessarily so. Deltaic flats are an obvious example, where the surface form is clearly dictated by the mode of deposition. Glaciers leave characteristic moraines of various forms, and these forms persist essentially unchanged for tens of thousands of years and longer. Sometimes the climate changes and a glacier melts away, following a period of no or minimal flow. Such an event produces very characteristic 'dead-ice' topography which is seen well preserved today, even though it is thousands of years old.

On closer examination we find, as is often the case in the studies of natural systems, that the easy classification, into erosional and depositional forms, is in fact not entirely satisfactory. Many fluvioglacial features, forms associated with the streams of melt water within or flowing from glaciers, are at once depositional since they are deposited stream sediments, and erosional because they are cut into terrace forms or similar by the streams themselves shortly after deposition. The landslide or mudflow on a slope is not normally defined as a feature either erosional or depositional. Yet the form of the moved mass after the event is quite characteristic, and entirely due to the actual nature of the movement. It is also 'deposited' – but not by an agent such as a river or glacier.

While we perceive denudation as an overall process of lowering, there are many pauses or relatively stationary situations. The formative agency, if a river which has long since migrated or cut deeper into its valley, or a glacier which has long since melted away, can only be deduced. The stationary situation clearly can persist for long periods, during which there may be radical changes in the environment leading eventually to renewed movement.

The process of denudation being therefore a continuing one it is appropriate to ask why the land surfaces have not become nearly flat in view of the great age of the earth. In fact the elevation of the land masses is repeatedly renewed by processes of uplift. That such uplift of the earth's crust occurs is, of course, demonstrated by the occurrence well above sea level of marine sediments and fossils, and also of igneous rocks originating at some depth within the earth.

The geomorphologists, W. M. Davis in particular in a series of papers (Davis 1963) some seventy-five years ago, developed the idea of uplift occurring periodically as the land surface approached a uniformly flattened state. The idea was not original to him, but he laid great emphasis on the repetitive nature of the cycles of erosion followed by uplift, and renewed erosion. It is only much more recently, particularly in the last twenty years, that considerable evidence has been obtained as to the nature of processes that produce uplift. It is now widely believed that the earth's continents are in slow lateral movement, carried by more or less rigid tectonic plates, perhaps 100 km thick, floating on slowly churning, warmer rock material. This churning motion follows from convection, a process where warmer material deeper in the earth rises because of its lower density and resulting bouyancy. This is illustrated in Fig. 1.1b, by an analogy. Convection must also include a counter movement in which cooler, heavier (more dense) material moves downwards. As the plates of much cooler, more rigid material move over the surface they may ride over one another to produce uplift. Uplift together with extrusion of materials also occurs above the rising convection current (Fig. 1.1a). An elementary account of these complex questions is given by Wyllie (1976), from whose book Fig. 1.1 is taken.

There is an additional aspect of these studies, worthy of note, because it has a parallel to certain smaller scale and much more local phenomena analysed in this book. Much of the deep-lying material which is warm and involved in the convective movements, is not in fact liquid. How then is it able to flow? In fact, the movements are extremely slow. We have a time scale different to that familiar in everyday experience. Over sufficiently long periods of time deformation of apparently solid materials can occur under quite modest stresses, this being particularly true when the temperature of the material is quite high, approaching although still less than the melting point. The source of the stresses are of course, the thermal energy in the earth's interior, and gravitation. In later chapters we shall see that, for example, the movements of earth materials on slopes, or the formation of ice masses in soils, involve knowledge of material behaviour of a kind that is quite different to that suggested by everyday experience or 'common sense'. Such forms of behaviour are only revealed by long and painstaking scientific study and experimentation.

In addition to the processes of uplift described there is abundant evidence of more

(a) **(b)**

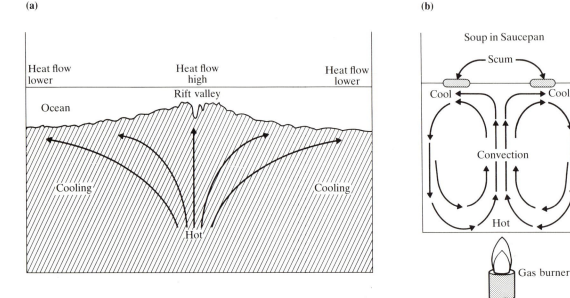

Fig. 1.1 (a) Vertical cross-section through a submarine mountain range illustrating schematically the slow convection of hot solid rock material rising beneath the central rift valley. (After Wyllie 1976)
(b) Convection cells in a pan of soup with floating scum on the surface. (After Wyllie 1976)

local adjustment of the elevation of land masses. Continental glaciation withdrew water from the oceans thus lowering sea levels and the weight of ice depressed the land; this is now rising due to the disappearance of the ice sheets.

The major processes which have produced the surface of the earth as we find it today are continually activated by the energy received from the sun, and by the earth's internal heat. For the purpose of study they may be considered as groups of interacting processes, each of which relates to a particular facet of the global environment. An example of such a scheme is given in Fig. 1.2. The term cycle appears frequently and refers to the sequential and repetitive nature of events. 'Rock cycle' for example (Fig. 1.3) is an element of the geological cycle, and traces the passage of rock and rock mineral material on the earth's surface and between the surface and the interior. The terms igneous, metamorphic and sedimentary are also applied to rocks as they are found at the earth's surface, and are described further in Chapter 2.

'Weathering' refers to the breakdown of rocks at the earth's surface and is discussed in the next chapter. The hydrologic cycle refers to

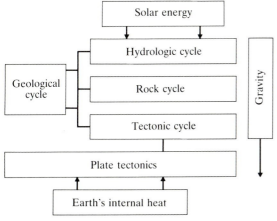

Fig. 1.2 Diagram illustrating interrelationship of major global processes. (After Wyllie 1976)

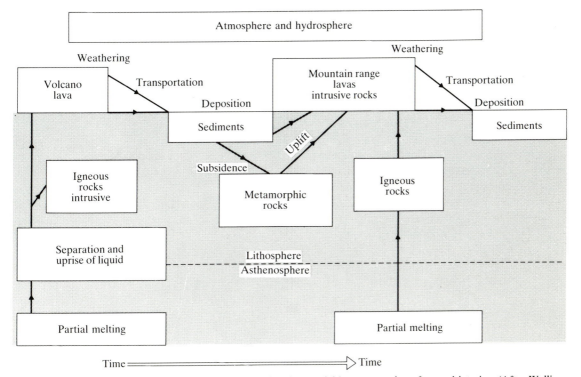

Fig. 1.3 The rock cycle illustrating continuing passage of rock material between earth surface and interior. (After Wyllie 1976)

the passage of water through liquid, vapour, and solid phases, and the exchange of water substance within and between the earth and atmosphere. The behaviour of water substance in the vicinity of the earth's surface enters into every chapter in this book. The hydrologic cycle, with particular regard to the ground surface is considered in Chapter 3. The all-pervasive importance of gravity is also indicated in Fig. 1.2.

The term 'cycle', as for example in 'hydrologic cycle', is not to be understood in too narrow a sense. We refer to a hydrologic cycle both with reference to conditions within a small area of the earth's surface and over a relatively short space of time, and also, with somewhat different meaning, to include exchange of water substance through geological time, with the earth's interior.

An understanding of the general nature of the landforming processes is basic to modern geology. Although such matters sometimes seem far removed from the everyday concerns of the applied earth sciences, without such a framework of knowledge we are inevitably limited in our understanding of the more immediate and local issues. Our knowledge for example of the origins of earth materials, of their behaviour in slopes, of the mechanisms of soil and rock breakdown, of climatic change, and of the behaviour of water below the earth's surface, has in large measure been built upon an understanding of the history of the earth's surface. Ignorance of such matters has sometimes proved costly. In the development for example, of the subject of soil mechanics, engineers were not infrequently guilty of overlooking the geological history and origin of different soil materials, and the major and minor landforms associated with geological events. In so doing, valuable clues were missed.

In trying to solve the more intractable engineering problems, such as those of the landslides in clays of marine origin which plague southern Scandinavia and the St. Lawrence-Ottawa river regions of Canada, often with loss of life, a detailed knowledge of the origin and history of the clays has proved a

necessity. Salts occuring in varying small amounts in these clays of marine origin have an important effect on their strength. The varying amounts of salts can only be understood in terms of the geological history, of land uplift and climatic change.

This section has merely introduced the geomorphological framework, within which our studies of material properties are pursued. There are many excellent texts in the field of general geology, some of which are listed in the bibliography to this chapter, and to which the reader is referred.

1.4 The interpretation of climate and weather

What is climate? Most people would answer rainfall, wind, air temperature, the variation of these with time and place. Common to these phenomena is a close, usually causal, relationship to the earth's situation and motion, relative to the sun. Could an inert mass, a 'dead' star or planet perhaps, in outer space, devoid of an atmosphere be said to have a climate? The body might have a temperature (although this might approach absolute zero). Would this be enough to justify the use of the word 'climate'?

The question is academic. We are concerned with the earth, and the steady supply of solar energy coupled with planetary motion gives immensely complicated climatic situations. We cannot possibly consider the physical form and dynamic nature of the earth's surface regions without the most careful examination of climate, and on a finer scale, of microclimatology – climate viewed as a local phenomenon. It is the exchanges of energy and mass at the ground surface, and their variations, even within a square metre of area, that must be examined if we wish ultimately to analyse the behaviour of the solid and liquid materials of the ground. It is suprising that 'physical geology' taught as a part of geology, has conventionally not included microclimatology.

Physical geographers have made many attempts to relate surface features to climatic regions. If these are broadly drawn, in terms of simple parameters of a qualitative nature: 'warm summers, cold winters, moderate rainfall' the exercise is unsuccessful. In a section of say, northern Canada, one might find treeless alpine-type mountain topography, adjacent to flat areas of muskeg (peatlands) with or without trees, or treeless tundra, or other quite distinct terrains. We might also find irregularly distributed 'permanently' frozen ground (permafrost). What can we say of the climatic controls? Trees require mean *summer* temperatures greater than 10 °C. Permafrost is associated with mean *annual* temperatures lower than 0 °C. Since the existence of permafrost is not inimical to at least some tree species, forest may or may not occur in association with permafrost. Trees and permafrost are associated with two *different* temperature parameters, which vary substantially independently. The problem is compounded by the fact that the mean annual temperature in question is not that of the air, which is the one we usually think of, but of the ground. This in turn is related to the air temperature only in a complex manner, depending on the precise nature of the surface cover, the vegetation and soil type. Thus attempts to classify climate in terms of its regional effects are made very difficult by the infinite number of combinations of individual climatic and other parameters and processes we may detect as being of importance. The effects of climate are modified greatly by topographic form and the nature of the ground materials. It follows that while climate is an immensely important factor in the behaviour of the surface materials, the effects of climate will vary with the nature and form of the ground surface and sub-surface materials.

Just as the Quaternary glaciation left its imprint on a large area of land, an imprint usually, but not always, quickly recognised by the geomorphologist, so do prolonged periods

of high temperatures and dryness, or other combinations of climatic conditions. There are a number of texts dealing with climatically defined regions, such as the recent work of Cooke and Warren (1973), on geomorphology of deserts, or that of French (1976) on the periglacial regions. Yet it is the diversity of the terrain rather than the uniformity of climate that provides the theme of such works.

Some further discussion concerning the nature of climatological data is necessary. Firstly, particularly in microclimatological considerations it is important to have a precise definition of the quantity under discussion. For example 'mean summer temperature' is itself open to objection. How long is summer? Are the means based on once-daily readings (a highly objectionable procedure) or on 4-hourly readings through every 24 hours, or perhaps on an averaging of somebody else's mean monthly temperature? How many years' observations have been involved?

Needless to say for any given place quite a range of possible 'mean values' could be obtained. This fact is well known, for example, to tourism promoters, who can usually manage to produce say, a pleasant mean November daily temperature for the most unlikely November holiday spots. Tourists are after all only outside in the daytime, so what more natural than to quote average noon temperatures? The fact that something like half of the days are at midday likely to be below the 'acceptable' mean draws attention to the variability – to weather as opposed to climate.

In the latter respect, the problem is not so much one of dishonest or careless manipulation of data, but of asking whether mean values are those that are relevant to the process or event under study. There are abundant examples illustrating that considerably more geomorphological change is often brought about in the course of a single exceptional weather event, than over tens or hundreds of years of more normal weather. The floods of East Anglia in Britain in 1953 are an extreme example, but on a smaller scale Rapp (1960) demonstrated how a single storm apparently caused the movement of more soil and rock material down slopes in Arctic Lappland than otherwise occurred throughout a 13-year period.

In predicting the likelihood of events, and their magnitude, for example predictions concerning landslides, one is often concerned to know the probability of relatively extreme weather events during the time of interest. To observe that the slope is stable in 'normal' weather tells us little. Even the ability to calculate what kind and amount of rainfall would cause a slide is of little use if the *probability* of such weather occurring is not known. This kind of problem is compounded for remote areas, by lack of climatological records, let alone the kind of detailed information necessary for full analysis.

As we are concerned with earth materials one final point needs to be made. Normally, when people speak of climate they are really referring to the climate in the air around them. A distinction is made between atmospheric meteorology which, broadly speaking, is what the airplane pilot is concerned with, and the 'Climate near the Ground' (the title of Geiger's classic work in microclimatology). But in studying soils and rocks we are concerned with another region just within the ground surface. The climate within the ground is different from that above; there is no wind, no rainfall, and the temperatures at a given time are not usually those occurring in the air. Other 'climate' processes dominate; heat conduction assumes great importance, as may the carrying of heat in slowly permeating water. Because only a small percentage of the sun's energy actually penetrates the ground, the climate within the ground usually receives scant attention in texts on microclimatology. Yet in the study of the behaviour of earth materials it is this climate with which we are ultimately concerned.

1.5 The role of biological activity

There is an elementary and erroneous view that sees 'natural' vegetation and animal life as being superimposed on a landscape whose nature and form has been essentially determined by inanimate, inorganic processes. A more advanced view is of plant and animal communities highly dependent upon and modified by the physical characteristics of their environments. But this view too misses one of the main thrusts of modern ecology. The science of ecology concerns the study of the relationships of living organisms to their environment. The presence and nature of vegetation always modifies the habitat, and biological activity may initiate drastic changes in the non-living environment. This is illustrated, in a world perspective, over geological time, in Fig. 1.4.

In contrast to the notion of the passive role of living things, is, of course, the popular conception of man as the great despoiler of nature. There is growing concern that through the extent and nature of his activities, man may initiate disastrous changes in his environment.

Worldwide industrial activity involves a rate of energy dissipation, or power, of some 10^{13} W. The energy of course, is currently mainly drawn from fossil fuels, coal and oil (ultimately the products of solar energy), from hydro-electricity (solar and gravitational energy), and to a limited extent from nuclear energy – the utilisation of which results in a rather large input of 'waste' heat into the earth's surface region. This dissipation, or utilization, of energy compared with the average 10^{16} W rate of arrival of solar energy at the earth's surface, may not seem too significant. But the world's climates are the expression of an immensely complex dynamic system in which an apparently minor change can lead to far reaching consequences. A rise of less than 3 °C in world average temperature would probably melt the world's drift ice (Lockwood 1978) which could be beneficial. But it might also start melting of the ice caps sufficient to cause inundation of many densely populated coastal areas as sea levels rose (Gribbin 1976). New desert areas would form (Bryson and Murray 1977); elsewhere agricultural conditions might improve. It is currently accepted that a general temperature

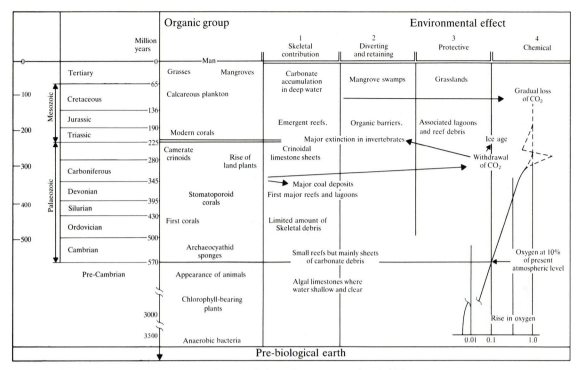

Fig. 1.4 The effect of some groups of organisms on their environments. (After Goldring 1971)

rise is occurring because of the increase of carbon dioxide in the atmosphere from industrial activity. The carbon dioxide acts somewhat like the glass of a greenhouse in keeping radiation in. If the combustion of fossil fuels continues to increase at current rates, temperatures are likely to rise by 2.5 °C or more within the next century (Hansen et al. 1981).

Justifiable concern over a few dramatic and obvious issues must not lead us to overlook the general interdependence of the living and non-living worlds. The importance of the energy balance at the earth's surface has already been indicated. A close, detailed study of the ground and vegetation, and its variation over short times or distances, is required before we can hope to have a proper, quantitative understanding. Vegetation intercepts solar radiation. That radiation will be absorbed, reflected or otherwise transferred in different proportions, depending upon certain properties of the vegetation in question. Part of the energy goes to evaporation. The process of transpiration of water passing from the soil,

through the plant (where it is involved in fundamental physiological activity) and into the atmosphere by evaporation, is of key importance. The rates of this transfer too are substantially governed by the nature of the plant cover, the species involved, as well as by the day-to-day environmental conditions of the locality. The process by which plants utilize solar energy to fix carbon in the form of fats, proteins and carbohydrates, is called photosynthesis. Agricultural crops, it has been estimated, utilize 10^{13} W of solar radiation in this way (Von Arx 1974), and a larger quantity presumably applies for natural vegetation.

Transfers of water by plants obviously affect the hydrological conditions, and the water balance, as well as the energy balance in the ground. The effects of dead vegetation in providing mulch, which retards drying of the soil, is well known to farmers and gardeners. The above-ground parts of plants, by reducing the near surface wind speed, also retard evaporation from the surface. A thick mat of close-grown mosses, grasses, etc. or of dead vegetation, also represents a layer at the surface of different thermal conductivity and heat capacity to that of the soil, with important effects on the temperatures and heat flows in the soil.

One of the more easily understood roles of vegetation is the mechanical one. Roots give strength to soils, and frequently prevent or reduce movements of soil in the near-surface layers of the ground. The importance of vegetation is well known to those concerned with the practical problems of soil erosion.

Rather more complicated are the effects of chemical changes occurring in association with vegetation, and the organic compounds which form from plant remains (humus). The near-surface weathering which gives rise to the pedologists 'soil horizons' (and which are often regarded by agricultural scientists as constituting 'the soil', in a narrow sense) is markedly related to the type of vegetation, as well as to the climate and other environmental characteristics.

Figure 1.4 illustrates the gross changes that have taken place in the environment in association with changes in fauna and flora over geological time. These changes are associated as much with the evolutionary process and the differentiation of species, as with the constantly changing physical environment. But on a totally different time scale, even well within a human lifetime, drastic changes may occur in the vegetation of an area. It is axiomatic in plant ecology that the vegetation of a particular site tends to follow a sequential pattern of development with time; the so-called plant succession. Thus an area of bare soil left by a retreating glacier is first colonized by small plants which thrive

on material which is merely shattered rock. As physical and chemical weathering proceeds, aided by organic products of plant decomposition, larger plants such as grasses and sedges become abundant. Subsequently, and if the climate is suitable, shrubs and even forest may develop as the ultimate climax vegetation. Another, more rapid succession is that to be seen on derelict building sites, which in a few years become covered with weeds, garden 'strays' and ultimately shrubs and trees.

This sequence of events may be disturbed, under natural conditions, by for example, climatic change, or more local effects such as fire, a change of soil, drainage conditions and so on, but each such setback is merely the start of a new, perhaps different succession. Naturally as each succession leads to new, different aggregations of species (plant communities), these will have different effects on the earth materials and land-forming processes.

Though we might consider man and his constructions, and plant and animal life as consisting of 'earth materials' this book cannot be extended to consider biological systems in detail. But in our examination of the immediate earth surface processes, and the associated materials we cannot avoid considering the effects of the vegetation cover. Detailed consideration of the local effects of biological activity is usually a preliminary requirement for considerations of more global effects as outlined above. The details of the 'physical' environment can no more be considered in isolation from biological activity than the converse.

2

Materials and processes

2.1 Nature of earth materials

In colloquial usage rock refers to a hard, durable material, while soil is soft, granular and easily separated between the hands. The terms 'soil' and 'rock' are used by scientists in several ways, and have several definitions. However, the simple distinction above is a good starting point. In this book 'soil' will include sand, silt and clay, and will not be limited to a fertile surface layer.

Almost all rocks are assemblages of several minerals, and the mineral composition is a means of classifying rocks. Soils are largely composed of small pieces of rock but also include certain minerals that are not commonly found, or only found in an altered condition, in rocks.

Many rocks have their origin as granular sediments deposited in water. Such rocks are called *sedimentary* if they are essentially unaltered since deposition, and might be regarded as either soils or rocks, depending on the degree of compaction. If there is some change due to high temperature, or pressure, and chemical change, associated with a period spent at significant depth in the earth's crust, the rock is '*metamorphic*' – a metamorphosis of the granular material has occurred.

Another group of rocks, *igneous* rocks, is quite different, being an assemblage of minerals, usually crystalline, and formed by the solidification of liquid materials originating at depth in the earth. Granite is an example. The assemblages of minerals are not perfectly stable in a chemical sense. They interact to produce new compounds, but because they are in the solid state the process is extremely slow.

Near the surface of the earth liquid in the form of water is available, and in soils the water comes into abundant contact with the particles. Together with climate-induced heating and cooling and the biological activity of plants and animals, water facilitates physical and chemical change. It also introduces chemical compounds from the atmosphere. The combined effects result in a much faster breakdown of the minerals and constitute *weathering*.

The effect is usually less marked in most rocks than in soils, but in any case is highly dependent on time. In regions covered by ice during the Quaternary glaciation, almost all weathered material is shallow and postdates the retreat of the ice because the older surface layers were eroded away. In other parts of the world weathered material, developed over milions of years, extends hundreds of metres into the ground. When rocks are broken down into soils by weathering *in situ*, the material is a *residual soil*. However most soils are accumulations of broken and weathered rock deposited by water or ice, far from their place of origin.

Regardless of the origin of the soil the material nearest the surface and extending some 2 metres or so downwards, shows the effect of contemporary, or, geologically speaking, recent weathering. The soil horizons in this layer represent a distribution by depth of minerals and organic compounds, in a manner which is markedly dependent on the climate. Agronomists and pedologists restrict their use of the word soil to the material affected by this stratification.

The terms 'soil' and 'rock' are used in this book in the sense used by engineers, which is essentially that outlined initially. 'Clay', 'mud', 'loam', 'sand', 'gravel', 'peat' all refer to 'soils'. Some properties of soils are comparable to those of rocks, and frequent repetition of the phrase 'rocks and soils' is avoided by simply referring to soils. Normally it will be obvious whether or not the property or process under discussion could also be related to rock but in some cases, rock will be the subject of a separate discussion. However soils constitute the surface of the earth over greater areas than does exposed rock and so the particular characteristics of rock will receive somewhat less attention than those of soils.

Soils may consist of a mixture of grains, or particles, of many sizes, or of grains of a fairly uniform size. Perhaps boulders are the largest 'particles' in soils; gravel refers to particles of 'equivalent' diameter greater than about 2.0 mm (the figure may vary). Of course, the soil particles are rarely spherical, and the 'equivalent diameter' may relate to size of sieve openings, or settling velocities, when such analytical procedures are used (Baver, Gardner and Gardner 1972). Particles of diameter: 0.02–2.0 mm may be referred to as sand; those of 0.002–0.02 mm as silts. The latter term is sometimes used in a genetic sense, of material deposited by rivers, but the terms are used here exclusively with reference to size of particle. Clay particles are less than 0.002 mm in size. There are several different systems of size classification so that the limits used by different authors vary somewhat.

That these sizes of soil grains or particles range through several orders of magnitude is responsible for much of the great variation in the properties of soils. Numerical values of the majority of material properties, as for example, strength, heat capacity, weight per unit volume, all differ markedly with the size of constituent particles. Soils which are assemblages of various sized particles often have intermediate values between those of soils of more uniform grain size composition. A graph showing grain size composition is one of the commonest ways of characterizing a soil (Fig. 2.1).

Soils, as granular materials, are also characterized by their porous nature. A feature of considerable importance mechanically is the fact that the pressure in the water within a porous material is often different from that within the mineral particles or on the soil sample as a whole. Consider pebbles half filling a can. At any point the pebbles carry the weight of those above them. After some water is poured in, the pressure in the *water* between the pebbles at some point is simply dependent on the height (or head) of *water* above the point, and this is not changed even after a weight is placed on top of the pebbles.

But in other respects water in the soil is different from that standing alone in a container. The pores of most soils are sufficiently small to produce capillary effects in water, which can thus be drawn into a dry soil. The effects are quite complex and represent an interaction between the water and the solids

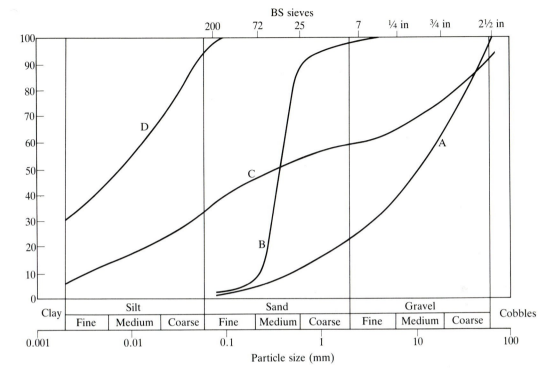

Fig. 2.1 Grain-size composition curves. For any point on a curve the ordinate represents the percentage of the weight of the whole sample, composed of particles smaller than the size represented by the abscissa. Curve (A) is for soil classified as well-graded sandy gravel. Curve (B) is for a poorly-graded sand, curve (C) is for a glacial till, a material categorized by having a range of grain sizes. Curve (D) is for a silty clay soil. (After Craig 1974)

significantly affecting the properties of the soil. The interactions between the surfaces of the particles and the water within the soil are challenging to the scientist. It is well known to physical chemists that water molecules, very near to a solid surface, are affected in various ways by that surface.

Forces of attraction emanate from the molecules of the solid; water molecules are held or adsorbed, and may be 'bound' to the molecules of the solid material, by a variety of physio-chemical effects. Not only are such water molecules less free to move than those in the bulk of the water mass, but water in the so-called adsorbed layer exhibits changes in a number of fundamental properties. Its density is changed, although there is argument as to how much; its viscosity is apparently changed; it is doubtful if we can properly speak of the pressure of such water. Pressure in water is normally understood as being a quantity equal

in all directions, but adsorbed water molecules are subject to forces acting essentially at right angles to the solid surface. It may be, for example, that an adsorbed water molecule can move *sideways*, that is parallel to the surface, more easily than it can move *away* from the solid surface. The freezing point of adsorbed water is a temperature lower than 0 °C, and the amount of heat to be removed to freeze it is less than for 'ordinary' water.

The effects of a solid surface as, for example, of soil minerals, or of the glass of an ordinary beaker, does not extend to more than about 10^{-5} mm from the surface. Consequently in considering the properties and behaviour of 'ordinary' water, we do not normally consider the adsorption effects of the container. After all in a beaker containing 250 cm^3 of water, only the minutest amount is actually within the range of the adsorption or 'surface' forces. The rest of the water is what we call 'free' or 'bulk' water. There is enough of it that its properties are not affected by its amount, or by the nature of the container.

In many soils, especially clay soils, the situation is very different, because the particles are so small there is a huge amount of surface. A gram of a clay consisting of the clay mineral illite may have 75 square metres (m²) of mineral surface. If the clay mineral were of the montmorillonite family the figure could be 750 m² – the area of the floor of a large lecture theatre. If one imagines the volume of the water from the pores of a gram of clay, being spread over the floor of the theatre it makes a very thin layer. All the water in the pores of such a clay is likely therefore to be close enough to the mineral surfaces to have its properties modified, compared to bulk or 'ordinary' water. This is significant in relation to the water itself, and also affects the way the water in the clay soil influences the strength and other properties of the soil.

In coarser grained soils, sands or silts, the grains are usually orders of magnitude larger; the *specific surface*, square metres per gram, $(m^2 \ g^{-1})$, are usually orders of magnitude less than in clays. To a substantial extent the water behaves as 'ordinary' water, and the behaviour of the soil is accordingly different. In many respects analysis of the properties is substantially easier, since it is not so often necessary to delve into sub-microscopic effects, effects which lie outside our normal or everyday laboratory experience of water.

A further matter of importance is that clays, or more specifically clay-sized particles are not generally composed of the same minerals as the coarser-grained soils. The latter can in fact be composed of a large number of minerals, of which quartz is the commonest. Some minerals are much more durable in sand or silt size particles than others. But although presumably quartz and other minerals can be ground finer, and can occur as clay-sized particles, most clays consist mainly of one or more of the 'clay minerals'.

Until this century there was considerable controversy as to whether 'clay' was in fact a particulate material. The argument was finally laid to rest by the electron microscope, which usually revealed an assemblage of 'flakes' (pl. 2.1), actually clay mineral crystals, often looking like soap flakes. The 'flat' shape is also in part responsible for the greater surface area as compared with the more nearly spherical shape of silt or sand grains. The flakes are sometimes arranged in a more or less parallel fashion, sometimes in apparently random fashion. The manner of arrangement has proved to be of significance in several respects, discussed in later chapters.

In addition to their distinctive shape and size in particles, the clay mineral crystals have a special structure. Layers of alumina (aluminium hydroxide) alternate with silica (silicon and oxygen). The arrangement of the molecules is such that other atoms and ions (especially iron and magnesium) may be incorporated into the crystal lattice, and such that there are bonding forces at the surface of the crystals (Baver, Gardner and Gardner 1972). The latter effects give rise to the cation exchange capacity, or the ability of the clay particles to attract or exchange ions with the soil solution (or liquid phase). Such phenomena, together with the large surface area of clays are extremely important with respect to the mechanical, physical and chemical properties of clay soils.

The clay minerals may be broadly subdivided according to the arrangement of layers in the crystal. The kaolin group of clays have a 1:1 layer structure, whereas the hydrous mica group (including chlorite and illite) have two silica sheets (the term sheet, implying a particular assemblage of atoms, is more specific than layer) alternating with one alumina sheet. 'Hydrous' refers to the fact that layers of water molecules occur between the sheets. The montmorillonite group also has a 2:1 structure, and the assemblages (or lattices) of the crystals expand or contract markedly with variation in amounts of water between the silica sheets. There is also a group of fibrous

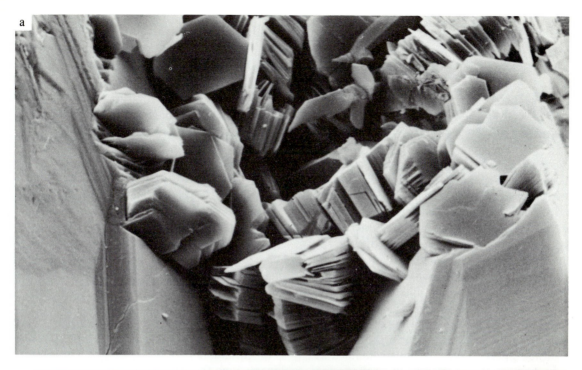

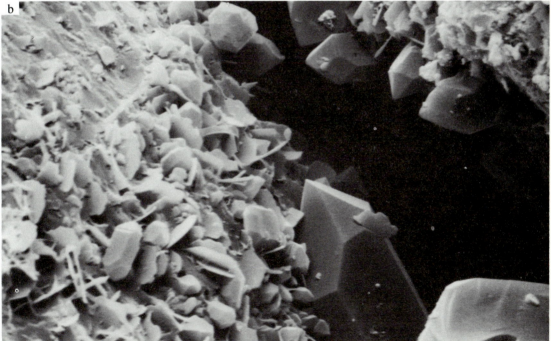

Plate 2.1 Photographs taken with scanning electron microscope of samples of clays. (a) Authigenic kaolinite (b) Authigenic chlorite and quartz. The particles of clay are so small that its particulate nature was only finally confirmed with the advent of electron microscopy (photos courtesy J. E. Gillott and the Journal of Microscopy).

clay minerals (palygorskite) which are porous, and can therefore absorb water (within the crystals) but without marked expansion.

2.2 Application of elementary mechanics and thermodynamics

The materials of the earth's surface behave in accordance with fundamental principles relating to energy and matter that have been thoroughly established by the work of physicists, chemists and others over several centuries. Interpretation of geomorphic processes requires that we apply such principles with precision and a proper degree of understanding of their meaning. Failure to do so results in theories or dogma, which are at best meaningless and at worst the source of danger to man's well-being.

Let us consider first a question of mechanics: of force and resistance. A force is defined as that tending to produce motion or change of uniform motion. Forces are ubiquitous and have several forms. They may for example act throughout a body, as do the effects of gravity, or they may be associated merely with the surface of a body where they in turn affect an adjacent body or particle. The mere existence of a force does not necessarily imply motion or change of motion. Consider a piece of ground. Below the surface there are forces due to the weight (a gravity effect) of overlying material. Most ground is not in motion (we ignore planetary motion) because the forces are counteracted by equal and opposite forces – the 'forces of reaction'. But, under appropriate conditions, motion does occur and then it follows that the 'forces of reaction', or resistance, are less than sufficient to prevent motion. Clearly, to predict whether movement is likely to occur, or to elucidate why it occurred, it is necessary to evaluate the forces involved.

Because weight is a force it is measured in newtons. We use the effects of weight in measuring quantities, that is, the mass (units: kilograms), of everyday things, and erroneously refer to the 'weight' e.g. ½ kg of foodstuff. Weight varies slightly with gravity at different places on earth. On the moon weights are greatly reduced. If it is understood that ½ kg refers to mass, then moon-travellers need not fear any 'loss' of their foodstore on arrival.

A force is the product of mass (kg) and acceleration. Acceleration is m s^{-2}. Thus force is kg m s^{-2} and 1 kg m s^{-2} = 1 newton. The acceleration, g, due to gravity (on earth) is ≃ 9.81 m sec^{-2}. Thus the weight of 0.5 kg (on earth) is 4.9 N (i.e. 0.5 times 9.81).

In the old metric system, weight in the proper sense was measured in kilogram-force (kgf), the definition being such that it corresponded numerically (in Paris, to be precise) to the kilogram mass, i.e. our kg in SI units. Using newtons rather than kgf avoids imprecision and possible confusion with kg.

Consider now the effects of weight in a uniform slope which for the sake of example, is of 'infinite' length (at least, a straight slope long enough that changes of form of the surface at distance have no effect on the point of interest). Assume also that it is free of water (because the presence of water complicates the analysis), and composed of at least fairly coarse granular material such as sand. Weight acts vertically, but vertical movement is hardly a possibility. Movement could occur logically only in a more or less downslope direction, and we should therefore evaluate the forces in that direction. To do so we must consider the weight of a vertical column of soil, extending from the surface to the depth Z of interest, and of horizontal cross-section 1 m^2. This weight is given by the unit weight times the height of the column. Unit weight, γ, is the weight (in newtons), per unit volume of soil, m^3, and the height of the column equals Z, the depth in the ground, in m. One can visualize the column as a pile of cubes (Fig. 2.2a). The expression γZ has the units N m^{-2}; this is often called Pascals: 1 N M m^{-2} = 1 Pa, especially in fluids.

N m^{-2} is not a unit of force but of a *stress*, which is force per unit area. It is usually more convenient to work with stresses as normally one is concerned with planes. Let us assume that the plane of interest is parallel to the slope surface, that is, movement would involve dislocation along such a plane (it can be shown that any other plane is less likely, under the conditions specified). Because γZ applies in a vertical direction on to a horizontal plane it is not the stress we ultimately require. Examination of Fig. 2.2a shows that the weight of the column is not bearing on 1 m², of the sloping plane AB but is spread over a greater area $\dfrac{1}{\cos \beta}$ 1 m². Thus our vertical stress becomes

$\gamma Z \cos \beta$ N m^{-2} – where β is the angle of slope

The force of interest is that in a direction downslope, along (parallel to) the plane of interest, AB. According to elementary mechanics, this force will be cos (90 – β) times that in the vertical direction (Fig. 2.2b). Consequently we have a stress, known as the *shear stress* τ, along AB:

$$\begin{aligned} \tau &= \gamma\, Z \cos \beta \cos (90 - \beta) \\ &= \gamma\, Z \cos \beta \sin \beta \end{aligned} \qquad [2.1]$$

It is referred to as a shear stress because it is acting towards producing a shearing of the soil on a plane parallel to itself.

We see that the force, and stress, tending to cause movement will be greater: (a) as the density, hence weight of the material is greater; (b) as the depth Z is greater; (c) as the angle of slope is greater (until the angle of slope exceeds 45 °).

(a) and (c) appear eminently logical. (b) might lead us to suppose that the risk of movement increases with depth. In the case of the dry straight slope of coarse-grained material this is not the case, because the strength of the soil also increases proportionally to the depth of material (Z) above the plane in question.

Strength is better considered as a property of the material of which the slope is composed, and detailed consideration is therefore deferred until Chapter 4. However, a general consideration is helpful. Calculating the stresses tending to cause movement does not tell us that

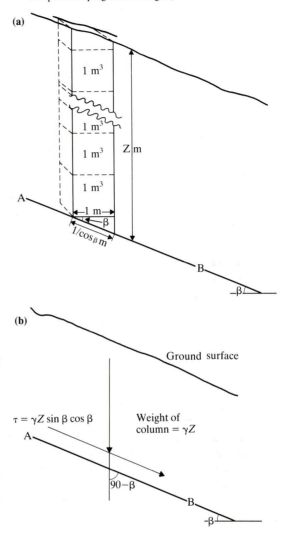

A column of soil, cross-section 1 m², bears on 1/cos β m², of a plane sloping with an angle β

Fig. 2.2 (a) Diagram illustrating vertical stress acting on sloping plane in simple slope (see text)
(b) Diagram illustrating the effect of weight in giving rise to shear stress τ (see text)

it will occur, nor what the strength, or ultimate resistance, of the material is. A resistance or reaction equal to $\gamma Z \sin \beta \cos \beta$ has already been developed, or the slope would not be standing. But the slope angle β could be greater, perhaps only by a minute amount, or perhaps substantially, before movement would occur. Thus the shear strength along the plane must be greater than $\gamma Z \sin \beta \cos \beta$.

Most slopes are neither uniform nor dry, and the analysis of the forces tending to cause movement is correspondingly more complex. The calculations may be such as to be conveniently carried out by computer. Nevertheless, the simple and 'ideal' case considered is very instructive, and indeed can quite often be applied to those real slopes where deviations from the 'ideal' are established as not being of overwhelming significance.

In the strict sense strength is something only achieved in the moment preceding breaking, rupture or deformation. When we speak of strength of an object we are normally referring to the maximum stress it can bear without breaking, and the same applies to earth materials. Strength has the units of stress, N m^{-2}.

Many materials show more than one kind of strength. For example, a certain stress may be sufficient to cause a very slow deformation which perhaps ultimately ceases without major dislocation, while a greater stress is necessary to cause a rapid, total rupture with consequent major displacements.

In the foregoing we have considered movement of material under certain specified conditions. There are other kinds of movement of mass, such as the flow of water through soil pores, which must be approached in a somewhat different manner. In our studies we are also concerned with flows of energy, especially flows of heat energy (or heat transfer as it is often called). In analysing the mechanics of slopes, the starting point was the existence of a force, due to gravity and therefore acting in a vertical direction. We then proceeded to calculate the resulting stresses, as they applied in the direction of likely movement. Because force always has direction it is a *vector* quantity. The same cannot be said for example, of pressure in a fluid, nor of temperature, electrical potential, energy or work, which are all *scalar* quantities.

Although these quantities are not specified by direction, we know that for example heat (thermal energy), flows from higher to lower temperatures, and electrical energy flows from points of higher electrical potential to those of lower. We say that heat flows along a temperature gradient. The rate of flow depends on the magnitude of the gradient i.e. the change of temperature per unit distance. The analysis of such situations is described basically by equations of the form:

$$q = k \, \nabla\phi \qquad\qquad [2.2]$$

where q = the rate of flow
k = the coefficent for flow (i.e. the conductivity of the material through which flow is occurring)
$\nabla\phi$ = the gradient of potential (or e.g. for heat, the temperature gradient)

Such an equation applies to the flow of gases, or of liquids, where the gradient of potential, is, for example, a pressure gradient.

The application of the equation is often complicated. A gradient may vary with time, or with distance. Suppose the coefficient k is constant along the flow path but that the gradient changes. Where the gradient changes, then the rate of flow q ([2.2]) must change. Now if q is different at different points along the path, whatever is flowing must be accumulating or depleting between those points. The accumulation of heat energy for example, causes a rise of temperature. In Chapters 5 and 10 we shall see how the distribution of temperature and heat within the ground, at different times, is related to such elementary principles.

Water has mass, and 'force' is defined as 'that which produces motion'. We have seen how the movement of soil on a slope can be analysed in terms of gravitational forces. Can we also understand the movement of water in soils in terms of forces, when our initial approach ([2.2]) is apparently so different? Indeed we can. When we speak of fluid pressure (and pressure differences, which give the pressure gradient) we are referring to force acting equally in all directions, at a given point. The pressure is often due to weight (a force), such being the *hydrostatic pressure* which increases with depth below the surface of a body of water. Because water has effectively no strength (except under certain circumstances not relevant here, see p. 77) the force appears not only vertically at a point, but in all other directions as well. Pressure is therefore a scalar quantity even though force, for example, the weight, is a vector quantity.

Gravity is not the only source of force acting on soil water. For example, as already mentioned, there are forces 'pulling' water molecules on to the surface of particles. These forces operate all round the particles and therefore also in a direction counter to that of eventual flow. But we can easily see that if these adsorption forces are more effective in one region of the soil mass than elsewhere, water will tend to move towards that region. It is convenient to speak of the *potential* of soil water, because the term allows us to refer to the effects of various forces, of which, individually, little may be known.

Movement of mass (e.g. water) constitutes work and a difference of potential between two points is a measure of the energy available to move the mass from the one point to the other. When we talk of potential, and energy, we usually mean potential and energy differences, or increments. In determining the difference in water potential between two points that we shall call C and D, in a system, two steps are involved.

Firstly, we must define a common datum, against which the potentials at C and D will be measured. For problems where simple considerations of pressure suffice, this might well be 'zero' pressure, which rather arbitrarily, we often consider to be the pressure of the air around us.

Secondly, we proceed to evaluate the potentials at C and D relative to the datum. In complex materials like soils, it is hard to do this by a theoretical process calculating and summing up the effects of the several physical phenomena producing forces. Instead we turn to experimental procedures. The essence of these is to bring the soil water into equilibrium with a body of pure water (in a bottle perhaps, with a connecting tube) the condition of which can be more simply described. Equilibrium means that there is no tendency towards movement of water between the soil and the connected body of water. The potentials of the soil water and the connected body of water are then equal. Suppose that we are considering a sample of the soil at C. To prevent water from moving into the soil it may be necessary to lower the pressure on the pure water, but, under appropriate experimental conditions nothing more is necessary. The potentials of the soil water and the pure water are now equal. Pressure has the dimension of force per unit area ($N\ m^{-2}$), and the pressure applied to the pure water is a measure of the potential (since the pressure is the only relevant source of potential in the *pure* water).

Repeating the experiment with a sample of soil from point D, gives the potential there. The difference if any between the potentials at C and D is now evident. In Chapter 6 the potential concept, and the sources of soil water potential are discussed in more detail. Once a potential difference, and the potential gradient has been determined, it remains to evaluate the coefficient k, known in this case as the hydraulic conductivity. This is a measure of the ease with which water flows through the material. Chapter 6 discusses these matters in detail, and should prove no more difficult of understanding than the basic concepts presented here. But the reader who wishes to strengthen his understanding of the fundamentals is referred to basic texts in physics and hydrodynamics.

The concept of a unifying quantity, such as potential, is extended further in the study of thermodynamics to include the effects of temperature. Heat is one form of energy, and in the study of processes at the earth's surface we are often concerned with heat and temperature and their relation to transfer of mass. Chemical thermodynamics includes the transformation of compounds in chemical reactions, or of the transfer between phases of a substance (water to vapour, or ice for example). In our studies we are particularly concerned with equilibrium thermodynamics – the study of equilibrium between different parts of a system, and disturbances of equlibria. A quantity known as the Gibbs free energy will be utilized. This is in many ways analogous to, and not infrequently equal to, the potential as discussed above with respect to soil water. But the fundamental definition of Gibbs free energy is that it is a mathematical function, of pressure, volume, internal energy, a quantity called entropy, and temperature. The Gibbs free energy G is defined by:

$$G \equiv E + PV - TS \qquad [2.3]$$
$$\equiv H - TS$$

where E = internal energy
$\quad S$ = entropy
P, V and T = pressure, volume and temperature
$\quad H$ = enthalpy or 'heat content'

Of greater utility are the derivations (see e.g. Nash 1970, p. 102).

$$dG = VdP - SdT$$
$$dG = dH - TdS$$

The latter equation is important in providing a basis for calculating the change of free energy ΔG *always* associated with a chemical reaction. A reaction occurs spontaneously if the free energy of the system of reactants and products *decreases* in association with the reaction, that is, if ΔG is negative.

A reaction can be endothermic (involving absorption of heat) or exothermic (releasing heat), and this is reflected in ΔH. Any spontaneous reaction involves an increase ΔS, in entropy. ΔG varies with pressure and temperature, and this explains why reactions occur spontaneously only under certain conditions of temperature and pressure. Again the reader interested in thermodynamic fundamentals is referred to basic texts (see also Ch. 6).

The Gibbs free energy can be thought of as a kind of common currency, which can be used to define equilibrium or to measure the states of disequilibria between the parts of a system, parts which otherwise can only be described by reference to a variety of apparently disparate properties.

The essence of thermodynamics is that it refers to macroscopic properties and is perfectly general. As discussed above with reference to potential, we can often avoid delving into the ultimate complex and microscopic, physical nature of our system. This does not mean that we have no interest in the microscopic nature and behaviour of earth materials, but that frequently we can arrive at highly significant relationships without first having to develop a very detailed perhaps currently unattainable knowledge of the system with which we are concerned. It is worth emphasizing that consideration of energy relations, and particularly of the Gibbs free energy (sometimes called the 'available' energy), gives insight into transfers from one chemical state or compound to another; from one phase to another; and into the movement of mass from one position to another. Chemical change is considered in the next section, particularly in relation to weathering.

There is a process by which mass is translocated, additional to the flows or movements already discussed. This is the process known as *diffusion*. If a spoonful of sugar is poured into a cup of water which is not stirred, after a period of time the sugar nevertheless will have dissolved and be present in uniform concentration through the water. Characteristic of such a diffusion process is that the random motion of molecules which occurs continuously, at the same time involves net movement towards regions of lower concentration. Another example is dispersion of water vapour of uniform temperature in a still atmosphere which takes place wherever there are differences of concentration (mass of vapour per unit volume of air), or, what is the same thing, when there are differences of the atmospheric vapour pressure (Ch. 6).

Diffusion processes of this kind are described by Fick's law, whose mathematical form is similar to [2.2] The conductivity coefficient is replaced by a diffusion coefficient whose value depends on the substances involved, and also on temperature and pressure. The gradient term becomes a gradient of concentration, measured as the mass of diffusing substance per unit volume of solvent, air, or whatever medium in which the diffusion is occurring.

Because diffusion involves spreading or dispersion of one substance in another it necessarily involves changing concentrations. Accordingly, to take account of the change of concentration gradients with time and distance somewhat more complex 'diffusion equations' are useful. Equations of this type will be briefly introduced in connection with several topics in this book. When used in studies of heat and temperature distribution (see Ch. 5) and moisture distribution in soils (Ch. 6), the phenomena involved are not those of diffusion in the sense of the word as used previously, that is, a process involving mixing of solids, liquids, or gases, due to the motion of individual molecules. In studying the thermal properties of soils, for example, a quantity known as 'diffusivity' will occur, but this refers to the dispersion of heat energy, and not to mass. Although thermal diffusivity has the same units of measurement, $m^2 s^{-1}$, as a diffusion coefficient, and may be used in similar mathematical operations, the two terms are not synonymous.

In concluding this introductory discussion to flows or fluxes of mass and energy a further term will be defined. Suppose that one has a uniform bar of metal the ends of which are maintained at different temperatures. Assume further that all other conditions around the bar are unchanging. There will be established a steady flow of heat along the bar. This is not an equilibrium condition because heat energy is flowing. But nothing will change within the limited volume under consideration unless there is some externally-induced change, for example, a change of the temperature at one end. So long as a change does not occur, we have a *steady state* condition.

In considering the processes occurring in earth materials we often assume steady state conditions. They are instructive in that they involve the simplest of mathematical analyses.

In reality, steady state conditions are at best transient and confined to a small region: changing conditions are after all, the very essence of geomorphic processes. Sometimes it is useful to consider a situation as a steady state, knowing full well this is not strictly the case. For example, a river which is neither eroding nor depositing material at a particular point is said to be in a steady state condition. In the chapters concerned with properties of earth materials we shall attempt to be more rigorous than this, although rather similar generalizations will appear in those later chapters analysing particular field situations.

2.3 Physical and chemical change: weathering

The weathering processes are largely the result of climatic phenomena, particularly temperature change and the availability of water, and also to circumstances such as the lower pressures existing in or near the surface relative to those at depth. Most rocks and many soils have earlier experienced great pressures, exerted by overlying material, and the denudation processes by which they reach the surface involves an unloading, often with important consequences for their mechanical properties.

'Weathering' thus refers to the complex and varied phenomena by which the materials at the earth's surface assume their characteristic composition, but the term has no clear, generally accepted definition. In the present context, physical breakdown ascribable to the direct effects of flowing water in rivers, or of ice in glaciers will be excluded, and changes associated with relieving of overburden pressure will be deferred to later chapters.

The term physical weathering is applied to the fracturing of rocks as the result of stresses resulting from 'physical' changes, such as thermal expansion and contraction, the expansion associated with freezing, or with the formation of crystals of various salts. The latter may well involve some chemical change, and the term 'physical weathering' might better be replaced by 'mechanical weathering' thus drawing attention to the importance of pressures and stresses in this kind of weathering. The process can occur even with small particles.

The coefficient of thermal expansion is normally expressed as the change of volume of a unit volume of material per °C of temperatue change. It varies from a few millionths to a few thousandths for earth materials. Rocks and larger soil particles which are composed of several minerals experience differential stresses on heating or cooling, because of the different expansions or contractions of the minerals. This may occur especially as a result of the differential absorption of thermal radiation and consequent differential heating of minerals in rocks exposed to the sun. Uniform materials may also experience thermal stresses as a result of uneven heating or cooling.

Early experiments by Griggs (1936) on various rocks showed that even very large temperature changes (by comparison with those commonly occurring at the earth's surface) only produced occasional cracking even after many cycles of heating and cooling. The relatively small significance of thermal expansion and contraction as a weathering mechanism is also confirmed by later experiments (Journaux, Coutard *et al.* 1974). The process may be hastened by particular climatic or weather conditions, including rainfall, by partial confinement of a rock or boulder, and by other situations. The mechanical properties of the rock, its strength and its susceptibility to 'fatigue' under repeated thermal stressing are also important factors. This and other forms of weathering are reviewed by Ollier (1975). They are also considered in relation to particular climatic conditions, in later chapters.

Ice occupies 9 per cent more volume than the water from which it is formed. Freezing cannot continue unless this volume expansion can take place. Instead the pressure rises and at the same time a lower temperature is required for

freezing to continue. Only a degree or two below the normal freezing point the pressure is very high and liable to rupture any confining rock material. These pressure – temperature relationships are considered further in Chapter 7. The shattering of rocks by ice formation is well documented. Recorded falls of rocks on highways and railway lines for example, show a springtime peak. The shattering occurs of course, during winter freezing but the rocks are held in place by the cementing action of the ice, only to fall when the ice thaws.

Such shattering is not as inevitable as might seem, however. Much water held in rocks is not totally confined. Either the water is able to move out of the 'confinement' as ice forms, or the ice itself deforms if the rock is stronger than the ice. Small quantities of water may exhibit supercooling, whereby no ice forms even though the temperature may fall several degrees below normal freezing temperature. The frequency of shattering thus depends on many factors, such as the frequency of freezing and thawing, the availability of water, and the internal structure, deformation properties and strength of the rock. Experimental studies into this (and indeed into most weathering) are hampered by the difficulty of time scale and correctly defining the many combinations of factors which occur in natural situations. The certainty that ice formation – 'frost action' as it is often called – is responsible for much weathering must be weighed against the frequent observation of rock forms that have remained essentially unaltered since they were shaped by glacier action thousands of years ago.

An additional process of frost action occurring in rocks and soils must be distinguished. It occurs particularly where there is a network of fine pores and involves migration of water to growing ice crystals. In soils large ice accumulations are formed in this way, the resulting volume increase being known as *frost heave*. The large volume increase, also discussed in Chapter 7, does not involve such high expansive pressures but is known to fracture 'soft' rocks such as chalk. Certain harder materials with an appropriate porous structure may also be fractured in this way.

A process similar to the formation of ice is the growth of crystals of salts. This may follow from solution in water and subsequent concentration by evaporation, to the point at which crystallization occurs. The crystals may also be the result of chemical reaction; sometimes bacteria have a role in these reactions. Weathering due to salt crystallization is commonest in warm, periodically dry areas, but occurs in certain rocks in temperate areas, and even in the Antarctic.

The process of wetting and drying also gives rise to stresses in soils and rocks. These stresses can be said to have their origin in the forces which exist in the microscopic layer immediately adjacent to a mineral surface, and which follow from the mineral molecular structure. Water (and also cations in solution) being strongly attracted and held to such surfaces, means that considerable swelling may occur on the addition of water. The thickness of these adsorbed water layers then tends to increase, producing intergranular swelling pressures. On the other hand loss of water, particularly by evaporation, ultimately results in a thinning of the adsorbed layers. The water remaining is that held most firmly to the mineral surfaces, and in turn the particles themselves may be drawn together. Cycles of wetting and drying give rise therefore to a push-pull alternation and a resulting cracking and flaking.

The processes of chemical reaction constituting *chemical weathering* are in several ways infinitely more important than the simple fracture and comminution association with physical weathering. It is chemical weathering which gives rise to the clay minerals, so important in determining the behaviour of soils. Chemical weathering too is responsible for the formation and accumulation of important ores such as those of iron and aluminium. The availability of plant nutrients from rock and soil materials depends on chemical weathering, and, in the absence of its products, the vegetation cover is extremely limited. Chemical weathering has for the most part been studied by agronomists and pedologists concerned with agricultural fertility, or by geologists concerned with ores and their location. But in recent years the significance of chemical weathering has been realised in relation to mechanical and other soil properties. Sometimes chemical weathering has important consequences even when the changes brought about are minor in a lithological sense and not easily detectable. Thus an understanding of chemical weathering has assumed importance in soils engineering, in questions of pollution and public health, and other areas. In fact chemical change is as fundamental a characteristic of the earth's surface region, as movements of material or the annual passage of the seasons.

Fundamentally, chemical weathering (as all chemical reactions) consists of the reactions that occur because of the *lack* of equilibrium between substances (Curtis 1976). A state of equilibrium means, on the other hand, that the substances present are in forms that are stable at the existing temperature and pressure conditions. A definition of equilibrium is that the Gibbs free energies of the reactants and products of a chemical reaction, are *equal*, (that is $\Delta G = 0$ for the reaction, in accordance with our earlier discussion of eqn 2.3). If they are not, transformations will tend to occur, towards the condition of minimum free energy, that is, so as to increase the substances of lower free energy at the expense of those of higher free energy. Equilibrium is associated with particular temperatures and pressures and because these conditions are always changing, the materials of the earth's surface are very rarely in a state of true equilibrium. Equilibrium is dependent on the concentrations of reactants and products of a chemical reaction, because the free energy of each component in the reaction depends on its concentration. Consequently the condition of equilibrium will normally occur when a certain proportion of the reactants have been changed to the products. This is illustrated in Fig. 2.3. In this figure the free energy is shown on the

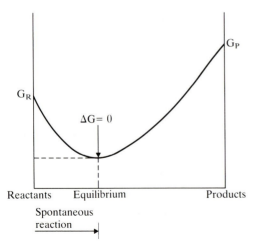

Fig. 2.3 Variation in free energy during a reaction.
(After Gymer 1973)

vertical axis. The horizontal axis refers to the progress of the reaction, so that the intercept of the curve on the left axis is the free energy of the pure reactants, while that on the right axis corresponds to pure products. The curve falls to a minimum – this being the equilibrium point where reactants and products have the same free energy, and the difference between them, ΔG, equals 0. It the products of a reaction are removed, however, the reaction continues. In this respect, the presence and movement of water is of extreme importance in chemical weathering. Clearly, in the case of weathering by simple solution the rate of weathering is determined by the rates of water flow and quantity of water involved. Water promotes chemical weathering in many ways.

One further general point needs to be stressed; the absence of equilibrium does not necessarily mean there will be immediate, measurable, change. In many cases chemical weathering occurs extremely slowly. In the absence of appropriate conditions for reactions to occur, a state of disequilibrium may persist over thousands, even millions of years. Studies of weathering must be as much concerned therefore with rates and conditions for change, as with the demonstration of equilibrium or non-equilibrium states (Fig. 2.4). In general, the rates of chemical reaction increase by two to three times for every 10 °C rise of

temperature (this following from Arrhenius' equation). In addition there are many reactions where, although the products have a lower free energy than the reactants, the reaction does not occur at a significant rate until there is an input of energy known as 'activation' energy. The rate at which such a given reaction occurs depends on the externally-imposed condition of temperature and other energy related factors. A fuller account of chemical reactions in relation to weathering is given in Davidson (1978) and references therein.

While the chemical changes involved in weathering are many and complex, several main categories can be distinguished. The most fundamental are hydration and the related processes of hydrolysis. The former relates to the ubiquitous attraction of mineral surfaces of water, already noted. This commonly involves the positively-charged poles of water molecules attaching themselves to points of negative electrical charge on the mineral, and various other forces are also involved. Hydration is a general term covering quite diverse effects. The tendency for water to be absorbed between the layers of certain clay mineral crystals is one example. The formation of gypsum crystals from anhydrite is another:

$$Ca\,SO_4 . 2H_2O \rightleftharpoons Ca\,SO_4 + 2H_2O$$
Gypsum   Anhydrite

Hydrolysis refers to a decomposition of mineral by water, involving the replacement of ions at the mineral surface characteristically by hydrogen ions from the water. Ions are charged particles released by dissociation of compounds, particularly in the presence of water. Water itself dissociates into positively charged hydrogen ions (H^+) and negative hydroxyl ions (OH^-). Other examples are given in Birkeland (1974).

An example of hydrolysis is the breakdown of orthoclase felspar to give the clay, kaolinite:

$$2KAl\,Si_3O_8 + 2H^+ + 9H_2O =$$
Orthoclase

$$H_4Al_2Si_2O_9 + 4H_4SiO_4 + 2K^+$$
Kaolinite

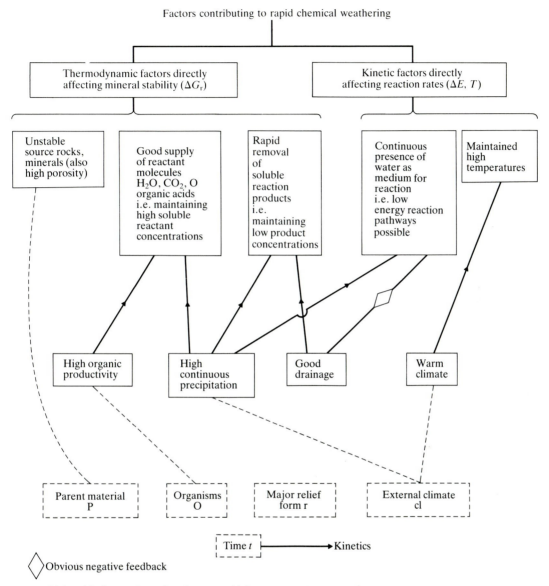

Fig. 2.4 Environmental factors which directly influence the nature, extent and rate of chemical weathering. (After Curtis, in Derbyshire 1976)

Most rock and soil minerals have positively charged cations such as those of sodium, potassium, and magnesium as components of the crystal structure. Because hydrogen ions are highly charged those in the soil water rather readily replace the metal cations, which are then transferred to the water perhaps to combine with the negatively charged OH⁻ ions.

The process may be continuous with the metallic ions moving out of the soil in solution. Alternatively a certain equilibrium concentration of cations may build up in the water (or, as it is often called, the soil solution). Minor changes, of temperature or pressure, or involving solution transfer, may then be sufficient to cause a recommencement of cation loss from the mineral. On the other hand there may be an ionic exchange in which similar, or different cations are again incorporated into the mineral crystal structure. The phenomenon of *cation exchange* refers to this aspect of hydrolysis in which one metallic cation may be replaced by another, and is of great importance. The importance is not limited to the grosser chemical changes associated with long-term weathering. Smaller disturbances to the equilibrium with respect to cation composition of soil solution and mineral, may have dramatic effects on the soil strength (Ch. 9), and are of significance in soil fertility.

The chemical reactions of oxidation and reduction are fundamental. They involve transfer of electrons to produce new assemblages of elements, that is, new compounds. Characteristically oxygen is involved, being carried in solution in the soil water while originating in the atmosphere, but many other reactions may occur to give different compounds. Iron compounds in particular are involved in oxidation and reduction processes. Reduction can be regarded as the reverse of oxidation, and occurs particularly in anaerobic (oxygen deficient) situations. In a chemical reaction, the oxidation of one component is associated with the reduction of another.

Chelation is a process whereby an ion, usually metallic, is drawn into the complex molecular structure of certain organic compounds. These compounds are known as chelating agents, and plants, for example, utilize them to extract ions from the soil as nutrients. Chelating agents are also used in laboratory techniques for removing ions from clays and soil solutions, for experimental purposes. Since chelating agents are produced naturally by vegetation and in organic residues it is likely that they have a significant role in cation exchange and other weathering processes.

An important control on many reactions is the concentration of hydrogen ions in solution, or the pH (which is the inverse logarithmic

measure of the concentration). Fuller accounts of this and other aspects of chemical weathering are given in Bunting (1967), Birkeland (1974), and Davidson (1978).

The products of weathering accumulate at the surface of the earth. In one sense all soils, that is all loose, particulate materials, are the products of weathering, since in a distant past they had their origin as igneous or volcanic material. But one generally considers sediments deposited in fresh or marine waters, or laid down from glaciers, as being virgin, unweathered, 'parent materials'. Weathering then embraces the changes which take place subsequent to such deposition. The quaternary glacial deposits are the parent materials for weathering which has proceeded since deglaciation; that is, often over a short period of some thousands, or tens of thousands, of years, and sometimes much less. The depth to which such weathering is conspicuous is at most some metres.

The importance of heat and moisture, that is, of weather and climate, is evident from considerations of chemical thermodynamics. The nature of the vegetation cover and the products of its decomposition are also important in chemical weathering. The products of weathering may, obviously, depend substantially on the material being weathered. The effect of all these factors is clearly seen in the differences in the *soil profile*, that is the characteristic sequences of weathered layers, or *soil horizons* from the surface to some limited distance downwards. The study of these profiles, and their variation from place to place or region to region, constitutes *pedology*.

Soil profile formation involves leaching, which is the transportation downwards of dissolved material (if upward flow occurs, the net transportation may also be upwards). Eluviation is the transport away of fine particles in moving water, while illuviation is the accumulation of particles by this means. Precipitation from solutions occur locally, when the moving solution becomes supersaturated with the solute. 'Organic accumulation' refers to humus and other products of vegetative decay.

Since the early days of pedology it has been realised that there is a general correlation between the nature of the profile and the different climatic and vegetational regions, or zones, of the world. Soils (that is, soil profiles) which fit well with this concept are known as *zonal* soils. But not infrequently some essentially local factor, perhaps poor drainage or a peculiar parent material causes substantial modification of the profile-forming processes, and produces an *intrazonal* soil. Yet again, there will be particular localities where for some reason the profile simply has not developed to the degree otherwise found (an example would be very recently exposed material). These incompletely developed profiles are *azonal*. In recent years classification of soils on this basis has been much criticized, particularly because it involves assumptions about the genesis of the soil, and overemphasis on the 'zonal' climate and vegetation in that respect. The explanation for the particular features of 'intrazonal' soils lies in some aspect of the weathering process and profile formation, maybe some quite obscure condition of the environment. Newer classifications are based on composition and properties of the soil profile itself – properties which, in theory, should be fairly easy to accurately define and describe. The fact that at least two systems (FAO-UNESCO 1974; US Dept. Agric. 1960, 1967), each complex and involving many new terms, are currently vying for acceptance need not concern us. Certain soil names of long-standing (some of which are included in the new classification systems, although quite narrowly defined) have a general usage. The following examples illustrate the dependence of soil type on weathering process.

The *podzols* are characterized by downward leaching leading to a characteristic bleached layer, which lies below both the surface humus and a layer of mixed humus and mineral soil. Below the bleached layer, will be another humus-rich layer, followed by a layer rich in sesquioxides, before the unaltered parent material is reached. These soils occur typically in humid temperate regions.

Under steppe or prairie conditions, in semi-arid climates typically a thick dark layer develops, which overlies, in the *chernozem* soils, a layer of calcium carbonate concretions. The latter follows from the incomplete leaching of lime, because of the low rainfall and infiltration.

Other zonal soils are *brown earths* (occurring under slightly warmer temperate conditions than the podsols and where deciduous rather than coniferous forest predominates), *prairie* soils (where dryness causes grass land rather than forest), and *grumusols* (clayey soils, of tropical or temperate areas with marked wet seasons, grassland or savannah). *Laterite* profiles occur under tropical conditions and are characterized by a reddish, iron-oxide horizon, with usually several horizons which are mottled or light in colour. The total thickness of the iron oxide horizon may be very great. The iron oxide is precipitated from ground water and often produces a hard almost rock-like material. The effects of temperature are considered in Chapter 12, and for further details of soil types reference can be made to books on soil geography (e.g. Bunting 1967, FAO-UNESCO 1974).

It is only rarely that an actual weathering event has immediate consequences for man (an example is the fall of stones and boulders). But the importance of understanding weathering processes lies in the ability to predict the nature and extent of different kinds of weathered material, and in predicting the properties and behaviour of such material both in the near and in the long term.

As may be expected from the variety of processes involved, rates of weathering vary greatly. Simple solution of limestone rocks causes a removal equivalent to some 4 cm of ground surface per thousand years in Northern England although different limestones or different conditions may give figures different by an order of magnitude (Ollier 1975). Soil profile development is observable after some 5 – 25 years depending on circumstances, but a 'full' profile may require hundreds of years in a temperate climate, and somewhat less in a hot humid environment. As weathering proceeds the products accumulate in deeper profiles and the rate of weathering will tend to decrease. Reference is sometimes made to 'fully-developed' or 'mature' profiles as though some final, equilibrium state is ultimately reached. Concurrently with weathering processes, there are generally sporadic, or more or less continuous, movements of material downslope. As a result there is generally no question of an ultimately stable situation, but only, possibly, of a steady-state condition in which soil profile formation approximately balances loss of material from the upper part of the profile. In fact the constant pattern of change of weather and climate, seasonally or over longer periods, shows that even a steady state in any precise sense is rarely possible.

3

Transfers of heat and moisture

3.1 Exchanges of energy and of mass; the conservation principle

The energy continually reaching the earth from the sun is, as indicated in Chapter 1, dissipated in various ways. The amount and nature of radiation reaching the earth's solid surface varies diurnally and annually, and also with latitude, cloud cover and many other factors. With respect to the processes occurring at and near the ground surface, these variations are as important as is the existence of the incoming energy itself.

Radiation is characterized by its wavelength, and is conveniently considered as 'short-wave' or 'long-wave' according to the dominant wave lengths present. Concurrent with the arrival (and, partial reflection) of short-wave radiative energy directly or indirectly, from the sun, is an emission of long-wave radiative energy, from the earth's surface. There are also other fluxes of energy to or from the surface, involving heat exchanges with the moist air, and heat flows in the ground towards or from the surface as outlined in Chapter 2.

In considering the behaviour of the rocks and soils at or near the earth's surface, it is important to establish the incomings and outgoings of energy, and also those of mass, more especially water. The converse is equally true: understanding climate and climatic change is possible only if the relationship to the earth's surface materials is clearly established.

If we consider a plane, for example the ground surface, then at any time the fluxes of energy to and from this plane can be represented in an equation for the energy balance. A simple example is:

$$R = H + LE + G \qquad [3.1]$$

where R = net radiative flux to the plane (i.e. incoming minus outgoing radiation)

H = flux of 'sensible' heat transferred to or from the air

LE = latent heat flux (evaporation or condensation)

G = flux into or out of the ground

It may be noted that the idea of a plane is somewhat unreal, in that the ground surface is hardly ever 'plane' (indeed vegetation is part of the surface in this context). But the concept is useful and often completely acceptable, so long as the inherent simplification is borne in mind.

Such an equation expresses the fundamental principle, of *conservation of energy*. It is not possible for energy to 'disappear'. At any time the sum of the various fluxes to the plane must be always balanced by an equal sum of fluxes leaving the plane. The form of the energy may of course change. For example if radiant energy is utilized in evaporation, then that energy is transformed into the internal energy of the vapour (and in part is dissipated by the work of volume expansion). Radiant energy may also, for example, be converted into 'sensible heat' – that is heat which finds expression as a temperature change.

A similar water balance equation is written:

$$p = E + I + r \qquad [3.2]$$
where p = rainfall
      E = evaporation
      I = water penetrating into the ground (infiltration)
      r = run-off (drainage of water over the surface)

This equation demonstrates again the conservation principle, in this case the conservation of mass. A correct water balance equation accounts for all the water arriving and leaving from the plane. If instead of the fluxes to and from a plane, we consider a layer of some definite thickness, there will be an additional term to cover the rate of storage or loss, of water or heat in the layer. For example, it is often important to know the amount of water held in the near-surface soil, say that above the water table. In such a case, we are concerned not only with water entering or leaving the ground surface, but also with that water entering or leaving the lower side of the layer by exchange with the ground water. There may also be lateral flows within the ground, so that the problem becomes a three-dimensional one in which we consider a column

of soil with exchanges of water both vertically and laterally.

Sometimes the word 'balance' is used in a narrower sense to describe the quantity, or change of quantity, of energy or mass stored in a defined layer. Such usage is found, for example, in glaciology, where a glacier having a 'positive mass balance' is one which over a defined time period has increased its volume. If 'balance' is used in this way, the implication is that the storage or change of storage is that which serves to balance equations similar to [3.1] and [3.2] but relevant to a layer or column, and where all the fluxes have been correctly evaluated.

3.2 Thermal and hydrologic cycles

Equation [3.1] involves fluxes, the units for which are J m^{-2} s^{-1}. Strictly speaking they apply only for a moment in time, because the values of the fluxes vary with time due to the continual change, for example of wind speed and insolation. However we can also draw up similar equations in which the totals of the different fluxes throughout a period of time are used. In Fig. 3.1 the annual heat balance for the entire planet earth is represented in the upper part of the figure. The proportions would, of course, be different for any particular location on the earth. In the middle part of the figure the heat balance of the atmosphere is shown, and at the bottom the heat balance of the earth's surface. The blocks represent the heat quantities exchanged as a result of the fluxes, and as, for example, a component of direct solar radiation passes right through the atmosphere, this is shown by a line circumventing the atmospheric balance.

A further point is that the annual balance (Fig. 3.1) at the earth's surface shows no heat entering or leaving the ground (no ground heat flux). This is because over a period of a year the ground heat flux averages out essentially (although not exactly) to zero. If there were to be a net flux of heat uniformly into, or out of,

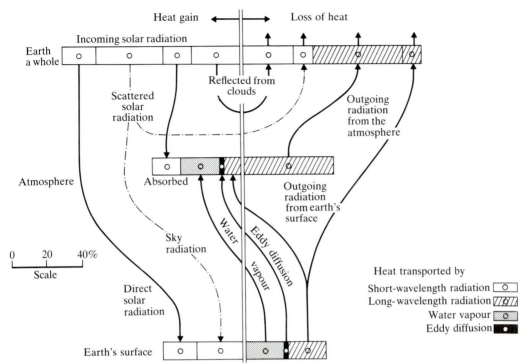

Fig. 3.1 Annual thermal balance of the planet earth, of its atmosphere, and of its surface. The size of the block indicates the magnitude of the annual fluxes. The arrows indicate their direction. (After Geiger 1965)

the earth's surface over many years, there would be a steady warming or cooling of the globe. At any particular site on the earth's surface temporary warming or cooling indeed occurs through the years, and the significance of this is discussed later. Quite different measurements also show that there is in fact a very small continual flux of heat from the earth's interior towards the surface (the geothermal flux, Chapter 1) but in the present context the effect is small enough to be ignored.

The diurnal heat balance, however, involves (on average) a net input or output of heat from the ground, because there is a warming or cooling trend according to the time of year. While ground temperatures vary diurnally and seasonally, the consistent pattern of ground temperatures from year to year reflects the existence of a certain quasi-equilibrium condition. The annual repetition of events is an example of a cycle – a term introduced in Chapter I and which is frequently used in

geomorphology to describe the passage of events in a periodic fashion such that at regular intervals almost identical states of the system prevail.

Figures for the four heat balance components of [3.1] summed for a whole year, for a site at Potsdam, Germany are given by Geiger (1965):

$$R = +8.2987\ 10^8; G = -7.58\ 10^2$$
$$LE = -8.384\ 10^8;$$
$$H = +1.62\ 10^3 \text{ J m}^{-2} \text{ yr}^{-1}$$

Clearly net incoming radiation (R) is substantially balanced by the evaporative component (LE). The reason the ground heat flux (G) and sensible heat (H) components are not closer to zero is that air and ground temperatures were not the same on the date of the beginning and end of the chosen year. These figures also illustrate the interrelationship of the thermal and hydrological balances, in that evaporation is dependent on the hydrological regime.

The term hydrologic cycle relates to the elementary concept that water passes from the atmosphere to the ground, as rainfall, snow or dew, and is subsequently re-evaporated, often after travel on or within the ground. The vapour in the atmosphere is then again the source of precipitation. The use of the word cycle in this context differs somewhat from that described with reference to heat, in that attention is drawn to the process of recirculation, or 'recycling' of water, rather than to the mere recurrence of particular states of the earth's atmosphere-surface system at uniform time intervals. The amount of water present in the world is essentially constant, but it is unevenly distributed and its distribution varies with time of year. Over a year the world's rainfall is approximately balanced by evaporation. Any exception would be due to increased or decreased storage within the ground, oceans, lakes etc. and considering the world as a whole this normally balances out to a small quantity. At a particular place, however, rainfall rarely equals evaporation whatever time period is considered. There is supply or loss of water by flow (surface or underground) from or to adjoining areas. If this were not the case, and if, for example, annual rainfall consistently exceeded evaporation, there would be a continual accumulation of water – a situation which common sense rejects. General experience tells us that the moisture conditions of the ground have a predictable annual constancy, even though varying with time of year. In this respect we have a hydrological cycle analogous to the thermal cycle.

Average annual precipitation decreases generally from the equator to the poles as does evaporation. In latitudes 10° to 40° evaporation is greater than precipitation although elsewhere it is usually less on an annual basis. Large-scale lateral redistributions of moisture must occur, largely by surface run-off and ocean currents, to provide the water for evaporation in the zone where this is in excess of precipitation. The processes are quite complex and are considered by Sellers (1965). For our purposes we are more concerned with the water balance as it applies to small and well-defined sites.

The water balance of the near surface layers, and the fluxes of water are of great importance to the natural processes occurring in those layers and at the ground surface. They control in large measure the mechanical properties and thus the denudation processes. Together with the interrelated fluxes of thermal energy, they determine the conditions for soil and rock weathering, that is for chemical and physical change, and also provide the conditions for plant growth. While cyclic passages of events with time are the norm, changes in the pattern of climate, or microclimate (brought about by change of ground surface conditions), are very important.

3.3 Local balances of heat and moisture

For our purposes we are primarily concerned with the heat balance defined for strictly local, and sufficiently small and uniform pieces of the ground. We shall see that knowledge of such heat balances enables us to understand the thermal regime, the distribution of heat and temperature in the ground, and the dependence on climate and ground surface conditions. In turn an understanding of the thermal regime is essential to explaining the behaviour and properties of the near surface layers of the ground.

Equation [3.1] is a simplification in that it groups the fluxes into the three basic processes of energy transfer: radiation, latent heat exchange and sensible heat transfer or conduction-convection ('convection' refers to heat transfer in association with movements of mass). It is necessary to analyse the role of the material components of the ground surface. For example, evaporation occurs directly from the surface of soil, but it also occurs by the process of transpiration from plants. Clearly it may for some purposes be useful to subdivide evaporation in this way, not least because the plant often exerts some control on its rate of transpiration. Whereas evaporation from a soil surface can be understood in relatively simple physical terms, evaporation by transpiration requires consideration of the more complex physiological processes of the living organism and in particular of the opening and closing of stomata which constitutes the plant's mechanism for exerting control on its own water balance.

Obviously, the nature of the ground surface, its moistness, whether vegetation or snow-covered and so on, will determine the magnitude of the components of the elementary balance [3.1]. The properties of different kinds of ground surface, with respect to the energy fluxes at the surface, will be the subject of Chapter 8. For the present the magnitude of the components of [3.1] will be considered in a general way by consideration of examples from various localities.

In Fig. 3.2a the annual passage is shown of the four principal components of the energy balance for a grass-covered site at Copenhagen, Denmark. The largest term at most times of the year is *R*, the net radiation. This represents all forms of radiation arriving at the ground surface less the quantity of outgoing radiation. At night, the net radiation is commonly a negative quantity but this is not evident when 24 hour, or longer, means are used as in this figure. During the winter months negative values also occur, and in fact all the terms are small, or become negative (again, as an average), including the flux in the ground *G*. The negative sign of *G* refers of course to the flux of heat within the ground *towards* the ground surface, that is, the cooling of the ground during the winter.

Figure 3.2b shows the components at a site at Flagstaff, Arizona. The net radiation, as might be expected from the angle of incidence of the sun's rays at low latitude, is larger. However, although solar radiation increases towards the equator, on a more local basis, this is not necessarily the case for net radiation. Deserts tend to have a high albedo, or reflectivity, such

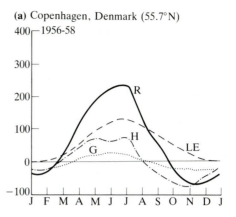

(a) Copenhagen, Denmark (55.7°N) 1956-58

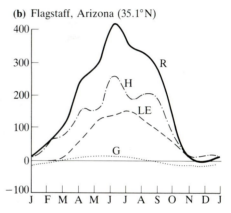

(b) Flagstaff, Arizona (35.1°N)

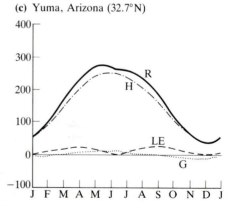

(c) Yuma, Arizona (32.7°N)

Fig. 3.2 The variation through the year of the average values for the component fluxes of the surface energy balance. Data for three locations. (After Sellers 1965) The values shown are in Langleys per day, one Langley per day being equivalent to 4.18×10^4 J m^{-2}

that the figure for net incoming radiation is reduced by the large amount that is reflected from the surface. This is seen in Fig. 3.2c where the clear skies at Yuma, Arizona are also important in allowing outgoing radiation to pass. The evaporation at Yuma is substantially less than at Copenhagen. This is because the dryness of the ground itself limits the amount of water available for evaporation. Net radiation is similarly small in polar regions on account of the high albedo of snow and ice, in spite of the large amount of radiation during the long summer days.

These few examples illustrate how significantly the magnitudes of the components of the energy balance vary from place to place. The variations are due on the one hand to gross factors of latitude and the relation of the earth to the sun, and on the other local, or microclimatic factors involving the nature of the ground surface.

The heat balance equation is normally presented (as in [3.1]) to show the disposition of net radiation. This is partly for the historical reason that early studies were made by atmospheric meteorologists, and partly because net radiation is, on land, commonly the largest component of the heat balance equation. But from our points of view it is the flux within the *ground, G*, which is of particular interest insofar as we are concerned with the thermal behaviour of the ground. The first thing that we note about G at any particular time (Fig. 3.2) is that it is generally small compared with the other fluxes. This circumstance means that a prediction of G from a knowledge of the other fluxes may be inaccurate. A relatively small error in the values for the other fluxes, results in a large percentage error in calculating G. It is also apparent that, for similar reasons, a change in ground surface conditions producing perhaps only a fairly small change in one or more of the other components may produce a rather large percentage change in the ground heat flux.

The magnitudes of the components of the annual heat balance are very easily changed. A disturbance of the surface, for example the stripping of vegetation, or the paving of ground with asphalt, results in substantial changes in the heat fluxes.

The thermal regime of the near surface layers of the ground to a depth of many metres, is determined by the manner in which the external or atmospheric climatic elements are modified by the complex exchange processes at the ground surface. In Chapter 8, the importance of the nature of the surface is analysed in more detail. An additional complication is that frequently gradual changes in the external, climatic elements are taking place over the years.

3.4 Climatic change

Historical geology is concerned with the conditions under which the rocks and sediments of the earth's crustal regions were formed. The nature of the sedimentary rocks and especially their fossil assemblages, are witness to the fact that the environmental conditions and particularly the climate have undergone repeated and drastic changes. There is evidence, for example, of glacial conditions existing hundreds of millions of years ago in areas which are today experiencing some of the warmest climates and conversely evidence of sub-tropical conditions are to be found in the rocks underlying the present-day polar areas. Evidence of continental drift and different orientation of the poles suggest that if we consider the sufficiently distant past, the entire disposition of the earth's surface features and climates was quite different. On a much shorter time scale, there is abundant evidence of the last ice age, which ended only some 10 000 years ago.

Yet until fairly recently there was a tendency to regard climate, when viewed on the time scale of, for example, a man's lifetime, as something essentially constant. Indeed, in the earlier parts of this chapter, the cyclic pattern of the passage of the seasons has been stressed, with its implication of variations of temperature or rainfall only within certain essentially constant limits, and about essentially constant mean values.

Plate 3.1 Mid-18th century moraine near Finse, Norway. The ridge in the centre of the illustration is a terminal moraine formed by a glacier which lay to its right – and retreated as a result of climatic change. The vegetation-covered surface to the left was last covered by ice thousands of years ago. At the time of the formation of the moraine, agricultural land was also overrun in the west of Norway.

When considering the behaviour of earth materials, and those naturally-occurring processes and events which are likely to be directly observed or experienced by man, we might assume that climatic change would be too long-term in its effect to be responsible for any dramatic turn of events or for sudden changes or reversals in the complex 'normal' patterns. The behaviour of glaciers within historical time is perhaps the most obvious evidence to the contrary (Pl. 3.1). Since the seventeenth century there has been a general decrease in the extent of the glaciers; for Europe this is documented in great detail. Often more than a kilometre distant from the ice front today lie glacial deposits laid down by the preceding advance in the seventeenth century. Recently there were signs that the general tendency for retreat was giving way to a re-advance, but climatic evidence contradicts this. These changes in glacier regime pose great problems in the utilization of glacier meltwater for hydroelectricity or irrigation, as the changing volume of the glacier ice leads to compensatory adjustments in the hydrological cycle and notably in the quantities of meltwater released in the summer. Both temperature and precipitation conditions affect the regime of glaciers. The intervention of glaciers in the hydrological cycle can result in more drastic consequences than the changes in temperature and precipitation themselves imply.

The extent of permafrost has in general, decreased during the last half-century and it is probable that during the last one or two hundred years, the permafrost has disappeared over areas of thousands of square kilometres. Disappearance of permafrost often leads to a special 'thermokarst' terrain (Chapter 11). The phenomenon together with that of glacial retreat is no doubt related to the well-documented rise of mean annual air temperature (Fig. 3.3) even though the rise at first sight seems small. To regard this correlation as a full explanation is, however, a simplification, not least because permafrost has

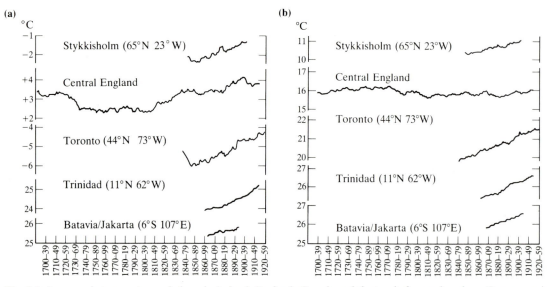

Fig. 3.3 Average air temperatures of places in Iceland, England, Canada and the tropical zone, based on 40 year running means (a) (above) for January, (b) (opposite) for July. Average of the two temperatures approximates the mean annual temperature. (After Lamb 1966)

actually increased in extent at certain locations (see Ch. 11). Obviously other, more local, factors must also be involved. It is important that a single parameter such as mean annual air temperature in no sense adequately describes a climatic change – it should be regarded as a symptom.

Climatic change represents a disturbance of a complex system, with modifications in circulation of air and water, re-distribution of atmospheric pressures, and thermal changes, often on a global basis. As yet, however, quantitative information is largely restricted to evidence of change in relatively few parameters. Climatic change is often promoted or otherwise modified by the nature of the ground surface, soils and vegetation, and the associated energy fluxes, that is, by the microclimate.

An important feature of climatic change is that it often involves an amplifying 'feed-back' effect. If, as noted in Chapter 1, a small rise in world temperature leads to a diminution of glacier and snow cover, there is increased absorption of solar radiation by the exposed ground surface. An element of solar radiation that was previously reflected back to space by

the snow and ice surface is now instead added to the heat energy stored in the earth's surface region, and the temperature increase is thereby augmented.

The effects of climatic change on agriculture and viniculture and the consequences for the human population are well-described in Ladurie (1972). Perhaps, as agriculture involves growing plants under conditions different from their natural habitat, it is natural that crops should be sensitive to weather conditions deviating from the average. But in many historical records it is possible to perceive trends, which are the repetition over a number of years of the same 'unusual' weather, and which must be described therefore as climatic change. When it has proved possible to quantify these trends, in the form of records of climatic parameters, they may appear small or even uncertain (Figs. 3.3, 3.4). The reason that along with such changes, there go quite dramatic effects in the disposition and behaviour of the surface materials, must surely lie in the sensitivity of the mass and energy balances, and the fact that an apparently small change in one or more components, may result in a much larger percentage change in another.

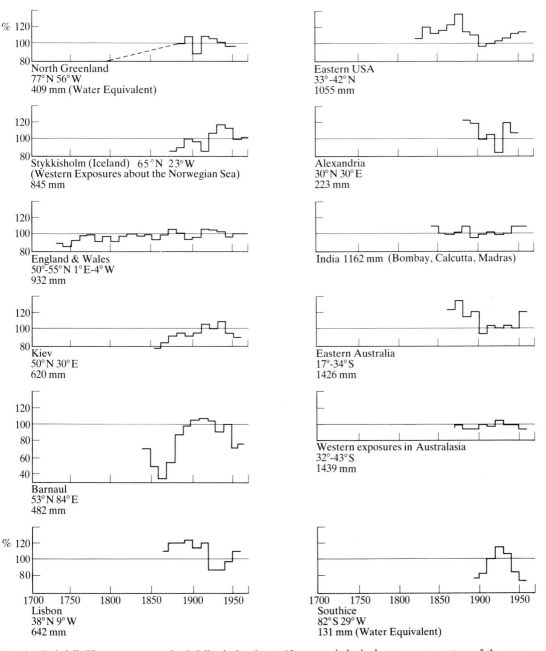

Fig. 3.4 Rainfall. The average annual rainfall calculated over 10-year periods, is shown as a percentage of the average annual rainfall in the period 1900 to 1939. The latter value is given under the name of each place. (After Lamb 1966)

Note again that the heat flux G ([3.1]) into or out of the materials at the earth's surface is small by comparison with the other fluxes, and is thus particularly liable to relatively large percentage change.

There are many theories as to the causes of climatic change. Suggested causes include variations in received solar radiation, due to variations in the earth's rotation and passage round the sun, or to disturbances of the earth's crust such as continental drift, or variations of sea level. Alternatively, the radiation passing through the earth's atmosphere may be modified (reflected, deflected or absorbed in part) in association with variations in dust concentration or in carbon dioxide quantity perhaps due to vegetational changes, or even contemporary industrial activity, as noted in Chapter 1. To some extent these and other postulated causes are interrelated, or even themselves the result as well as a cause of climatic variation. A good summary is given in Sellers (1965), and a more detailed discussion in Lamb (1972).

Until recently studies of climatic change were largely made for their intrinsic interest, and in connection with interpretation of geological or palaeobiological observations, or of the human historical or archaeological record. The significance of the disturbance of the ground surface energy and mass balance induced by climatic change, and the role of the ground surface and the near-surface earth materials in causing or modifying climatic change have received little attention (glacier regime being the notable exception).

This situation is likely to change. World food shortages require increasingly precise prognoses of agricultural conditions. The complexity of human society, of man's engineering and industrial undertakings, and an increasing appreciation of the interaction, for good or ill, of man with his environment, raise new issues. It is becoming clear that mankind, perhaps inadvertently, can himself cause climatic change (SMIC 1971). It is equally clear that our sophisticated civilization does not guarantee protection against disasters and distress of the kind that climatic change caused to earlier peoples, who variously starved as their crops failed, succumbed to pestilence and disease as other biological populations flourished and waned, and who struggled to overcome the difficulties of new and alien lands.

Modern science and technology also makes the human community vulnerable in other ways. For example oil and gas, a major industrial energy source, are increasingly transported great distances by pipelines. These are exposed to a wide range of natural conditions and risks. Where pipelines traverse cold regions the problems of providing safe foundations are increased by various effects, land sliding, ground settlement or expansion, associated with freezing and thawing (Williams 1979). Climatic change would increase these effects, and becomes a factor in the design of constructions expected to last more than a few years. In many areas even the extremes of weather to be expected are not fully known because observational records have only been kept for a relatively short time and the observing stations are widely spaced. The effects of these extremes on the properties and behaviour of the ground surface materials then become a matter of some speculation.

While the general climate clearly has a basic importance in the microclimate, the converse is not always the case, because of the broad 'mixing' ability of the air and wind. Thus the atmospheric climate shows a greater uniformity over distance than the microclimate which is subject to the variations of surface form, of earth materials, vegetation and of hydrological and other conditions. But when a change of surface or ground conditions is of sufficiently wide extent, it may have an effect on the general climate, including both the general thermal and hydrological balances for the region. The disastrous droughts of recent years in parts of Africa, and the spreading of deserts, are at least in part the result of agricultural or grazing practices modifying the microclimate over wide areas (UNESCO 1977 – discussed further in Chapter 12). In this book we are concerned primarily with microclimatological questions. It is likely that, in the future, more use will be made of detailed studies of the ground and ground surface, in explaining both causes and effects of climatic change.

PART II

Properties of earth materials

4

Mechanical properties

4.1 Mechanics and mechanical properties in relation to landforms

Mechanics is the branch of physics which deals with motion of matter, and with the forces that bring about motion. Motion, whether it be the flowing of water through soils, the falling of rain, the blowing of the wind, the slow movement of soil on a slope, or a sudden landslide, is characteristic of the earth's surface. Landforms are the product of movement of material; movement is therefore the fundamental geomorphological process. The desire to predict the nature, location, and magnitude of movements of soil and rock in the relatively near future motivates studies of soil mechanics in geotechnical engineering. The central aim of the geotechnical engineer is to influence such movements whether it is to prevent them, to minimize their effects, or maybe to accelerate them.

The analysis of soil movements requires an understanding of the forces involved, forces which are largely, but not exclusively, gravitational in origin. Gravity finds expression in weight, giving rise to stresses which may in turn cause movement. Ultimately the fundamental principles are similar whether we are concerned with the atmosphere, oceans or rivers, or soils and rocks. In this chapter those physical properties of soils are considered, which directly characterize their mechanical behaviour. It will be apparent that many of the considerations apply equally to 'hard' rock. The properties include, for example, density which determines the magnitude of forces a given volume may exert depending on its situation, on neighbouring material; and strength, which is the maximum resistance a material can develop against applied forces before deformation or movement occurs. The term 'mechanical property' is not always capable of rigorous definition. In addition to certain fundamental properties of earth materials, consideration is given in this chapter to the analysis of the mechanics of particular geometric configurations, that is the significance of slope, of the position of the water table and the effects of depth. The numerical values of many properties, the magnitude, for example, of strength, can only be evaluated when the surroundings of the material are known.

In analysing naturally occurring movements of soil or rock, knowledge of the chemical composition, and the origin or history of earth materials is often involved (see Ch. 9), although such characteristics are distinct from mechanical properties *per se*. Often, however, this additional knowledge enables us to predict mechanical properties, or explain the behaviour of larger masses than would ever be sampled or tested directly. Initially, we consider certain basic properties which are possessed by samples in isolation, although the field value may depend on the situation of the material.

4.2 Basic gravimetric and volumetric properties

Bulk density is the mass of unit volume of the soil, ρ, kg m^{-3}. By soil is understood the bulk material, including the mineral, water and eventual air or other components. The kilogram and the metre are both examples of *base* units in the SI system (see Davidson 1978). The gram and centimetre are accepted SI units, and g cm^{-3} for density may be more convenient on occasion. But there are good reasons for not using such sub-multiples or multiples (or those of other units) so they are generally avoided in this book. *Derivative* units involve a number of base units. In calculations concerning the more complex derivative units (for example, those for thermal conductivity: J s^{-1}°C^{-1} m m^{-2} = W m^{-1}°C^{-1}), it is easy to confuse say, cm and m, thus introducing an erroneous factor of 100. Furthermore, it is harder to remember ranges of values of material properties if various multiples of units occur, and it may also be difficult to visualize the relative importance of different terms in equations.

Bulk densities of soils commonly lie in the

range 1.5 to 2.1 10^3 kg m^{-3}. The minerals found in soils have a range of densities (see Terzaghi and Peck 1967, p. 27), but the dominant soil minerals, including clay minerals and quartz lie in the range 2.6 to 2.7 10^3 kg m^{-3}. An 'average' value of 2.65 10^3 kg m^{-3} is often used. For soils of unusual mineral composition a different value may apply. Water has a density of 1 10^3 kg m^{-3} and soil bulk density cannot exceed about 2.3 10^3 kg m^{-3} as this would imply too large a proportion of mineral material and too few pores to constitute a soil. Rocks are usually much less porous, and also because of the presence of heavier minerals have bulk densities which can be considerably greater.

Dry bulk density ρ_d is the mass of unit volume of a soil after drying (at approx. 105 °C). It is important to note that the unit volume is that of the soil *before* drying. Many soils decrease in volume on drying, while most rocks do not do so significantly. The dry bulk density is a measure of the compactness of the soil and is a property frequently required in calculations. Typical values are shown in Table 4.1. Because of the tendency of soils to change volume as water content changes, the dry bulk density is normally meaningful only if the water content is specified.

Table 4.1 Some typical values of dry bulk density ρ_d for several types of soil and rock

Soils and rock	ρ_d kg m^{-3}	
Uniform sand, loose	1.4	10^3
Uniform sand, dense	1.7	10^3
Mixed-grain sand, loose	1.6	10^3
Glacial till (many boulders)	2.1	10^3
Stiff clay	1.7	10^3
Soft organic clay	0.7	10^3
Very soft montmorillonite clay (high water content)	0.4	10^3

Density has particular significance in studies of soil mechanics, in that it determines weights. If density, ρ kg m^{-3}, is multiplied by the acceleration due to gravity g, 9.8 m s^{-2}, we obtain the weight per unit volume – the unit weight, γ, N m^{-3}. The unit weight is so frequently required in soils engineering, that 'γ' is commonly used in place of the expression 'ρg'. There is, theoretically, a loss of precision, in that g varies slightly with location on the earth's surface, and the value γ is thus not exactly a property of the material. For the engineer (at least if he stays on earth) the failure to distinguish the possibility of variations in g (the gravitational acceleration) is generally of no significance. In scientific studies, however, the use of 'γ' rather than the form 'ρg' may lead to serious confusion.

The definition of unit weight γ of a soil resembles that for bulk density, that for dry unit weight γ_d resembles that for dry bulk density, ρ_d, but with 'weight' replacing 'density' throughout.

Although the water content (w) may be expressed in various ways both gravimetrically and volumetrically, it is now defined internationally as the ratio of the weight of water to the weight of non-evaporable (at 105 °C) solids, expressed as a percentage.

Some clays, and also many organic soils, contain more water than solids, so that values greater than 100 per cent frequently occur. The city of Mexico is built on clays which may contain several hundred per cent water content by weight, thus as much as 10 times more water by volume, than solids. Such clays show remarkable mechanical behaviour and pose considerable problems to engineers (see p. 54).

Although the accepted definition of water content may appear curious, experience indicates it to be a good one both from a mechanical as well as a fundamental thermodynamic standpoint. Often, however, water content expressed volumetrically, volume of water per unit volume of wet soil, m^3 m^{-3}, is required in calculations. This is simply

$\gamma_d \dfrac{w}{\gamma_w 100}$ (–where γ_w is the unit weight of water,

N m^{-3}). Because of the great importance of specifying the form of measurement, the gravimetric water content should be referred to as w per cent *dry wt*. Dry unit weight is related to unit weight:

$$\gamma_d = \frac{\gamma}{1} \frac{100}{100 + w} \qquad [4.1]$$

where w = water content per cent dry wt.

The void ratio e is the volume of voids, that is, pores, cavities, cracks etc. whether or not waterfilled, expressed as a ratio to the volume of solids:

$$\frac{V_v}{V_s}$$

This is also a measure of compactness (or consolidation – the term used in foundation engineering).

The porosity, n, is similar except that the volume of voids is expressed as a percentage of the total volume. It is also expressed as a fraction, and is given by $1 - \dfrac{\rho_d}{\rho_s}$ where ρ_s is the average density of the solid components. If a soil is saturated, that is, contains no gas, then the porosity is equivalent to the volumetric water content.

The degree of saturation S_r is the volume of water expressed as a percentage of the volume of voids.

4.3 Soil strength

Strength means resistance to breaking or deformation, and is normally measured by applying force until breaking or deformation occurs. The force applied at that time is a measure of the strength. From a geomorphological point of view we are concerned sometimes with 'breakage' of materials, and sometimes with deformations.

In practice there is often some uncertainty as to precisely how much differential movement must take place before we can regard the strength as being exceeded. Very small movements, such as those associated with thermal expansion and contraction are not usually regarded as being matters involving strength of the material, but, as will be shown later there are a variety of deformations of limited extent which are of significance in the long term. For the moment, however, we restrict consideration to that strength, which if overcome is followed by an accelerating or at least a continuing motion. Strength of soils is commonly given as a *shear strength*, which is equal to and has the same units as that (shear) stress which acting parallel to the direction of movement produces such motion. Often the strength can be considered that value of applied stress, which, once exceeded, can then be followed by a somewhat lower applied stress without the motion ceasing.

Factors influencing the strength of different soil materials are discussed in the following sections. However, there is one important factor which is controlled not only by the nature of the soil at the zone of failure, but also by its situation. If one slides a book across a table, and then piles additional books on top, it becomes harder to push the book. The added weight constitutes a force acting *across* the plane of 'breakage' between the book and the table. The force can be expressed as a stress (force per unit area) which is known as a *normal stress* because it acts perpendicular to the plane.

The normal stress clearly affects the shear strength along the plane – which is also known as the plane of shear. Indeed, experiment shows that for many situations the shear strength, which in our example is the shearing force per unit area of book-table interface required to move the book, increases proportionally to the normal stress.

For dry coarse-grained soils there is virtually no shear strength in the absence of normal stress. The shear strength increases proportionally to an applied normal stress, and the rate of increase depends on a property of the material – a coefficient of *friction*. If we plot shear strength against normal stress applied, we obtain a graph as in Fig. 4.1. The slope of the line is tan ϕ. The angle ϕ is sometimes known as the *angle of internal friction*, but it is best regarded merely as the angle the tangent of which is the coefficient of proportionality of shear strength and normal

stress. This is expressed in the strength equation:

$$S = \sigma \tan \phi \quad \text{where } \sigma = \text{normal stress} \qquad [4.2]$$

applicable to dry materials having strength solely due to friction.

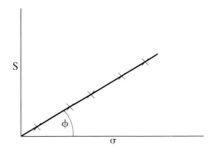

Fig. 4.1 Increase of shear strength S, with increase of normal stress, σ, for dry non-cohesive soil

Referring again to the ideal slope of Chapter 2.2, for which we obtained the expression for shear stress, we can quickly obtain the expression for the normal stress, σ, acting on the plane AB in Fig. 2.2b It is:

$$\sigma = \gamma \, Z \cos \beta \cos \beta \qquad [4.3]$$

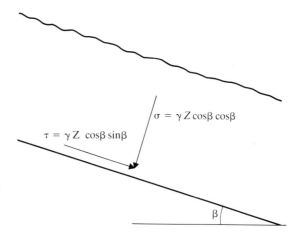

Fig. 4.2 Normal and shear stresses acting on plane parallel to ground surface, ideal slope

In Fig. 4.2 both normal and shear stresses on a plane parallel to the surface are shown. The special nature and situation of soils requires that the role of normal stress in relation to strength be examined further.

Water in the pores of soils has a highly important effect on the normal stress and thus on the strength. If an open-ended pipe is sunk into the ground water rises to a certain level in the pipe. At this level, called the water table, the water is clearly at atmospheric pressure (which is taken as zero pressure), while below this there is a positive pressure increasing proportionally to the depth h of the water in the pipe (Fig. 4.3). The pressure of the water, where the water table is horizontal, is $\rho_w gh = \gamma_w h$. The unit weight of water, γ_w, is 9810 N m^{-3}, and with h in metres, pressure has the units N m^{-2}. Assuming equilibrium has been reached between the water in the pipe and in the soil, then the same pressure occurs (at the same level) in the pores of the soil. It is then the *pore water pressure*, and usually given the symbol U. To distinguish pressure, a scalar quantity, from stress, a vector quantity, the units for pressure are often called pascals, 1 Pa = 1 N m^{-2}. Quite often the pore water pressure is referred to merely by the height, h, in metres.

If we consider the normal stress acting on a horizontal plane below the water table in level ground, the weight of water in the pores of the soil is included in the soil unit weight γ, and thus is included in the calculation of $\sigma = \gamma Z$. To make absolutely clear that in calculating the normal stress σ, we have added in the weight of water in this manner, we refer to it as the *total normal stress*. Z may also have to be subdivided into layers to allow for different moisture contents, and consequently different values of γ.

Any material submerged in water is subjected to a buoyancy effect, which reduces the apparent weight; to a value of zero in the case of a floating material. If we consider the normal stress as acting across particle to particle contacts it is apparent that when these are located in water the normal stress will be reduced. This 'reduced' normal stress is called the *effective normal stress* and is represented by σ. The use of the word effective refers to

the relevance of this value to the assessment of strength (again reflect on the books on the table, and the effects of repeating the experiment under water – with the table anchored). Submergence reduces the resistance to movement, the strength.

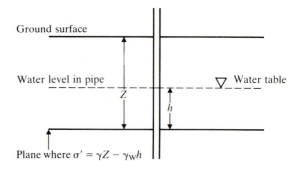

Fig. 4.3 A diagram illustrating the effective normal stress σ, and its relation to pore water pressure. The pore water pressure is revealed by the level of water in stand pipe

The distinction between 'total' and 'effective' normal stress is extremely important. Evaluation of the effective stress is almost always required in analyzing the strength of damp or wet soils. In the simple case (Fig. 4.3) of the horizontal plane at depth Z, at a distance h below the water table, in level ground, the effective normal stress is:

$$\sigma = \gamma Z - \gamma_w h \qquad [4.4]$$

*The term γZ is the total normal stress σ (again it may be necessary to take into account the variable nature of γ), and the term $\gamma_w h$ is the pore water pressure, U.

Calculation of σ and U is not usually as simple as in this example, but the relationship:

$$\acute{\sigma} = \sigma - U \qquad [4.5]$$

is of general validity and constitutes a fundamental tenet of soil mechanics. The relationship was established and its scientific and practical significance was first realised by Terzaghi (Skempton 1960).

In [4.3] total normal stress was shown for a uniform dry slope: $\sigma = \gamma Z \cos \beta \cos \beta$. If, instead, water occurred with a water-table parallel to the surface at a vertical distance h above the plane of interest, then the effective normal stress is:

$$\sigma = (\gamma Z - \gamma_w h) \cos \beta \cos \beta \qquad [4.6]$$

For the case of water filled pores equation 4.2 is replaced by:

$$S = \sigma \tan \phi \qquad [4.7]$$

This is the equation of strength for saturated coarse-grained soils. A more complicated situation occurs in fine-grained soils, and is considered in section 4.6.

4.4 Measuring the strength of soils

The simplest experimental apparatus for measurement of shear strength is the shear box (Fig. 4.4). The soil sample in the box is caused to shear along the plane between the two halves, by application of sufficient force P. The shear strength is:

$$S = \tau_s = \frac{P}{\text{Area of shear plane}} \qquad [4.8]$$

where τ_s = shear stress when shear occurs.

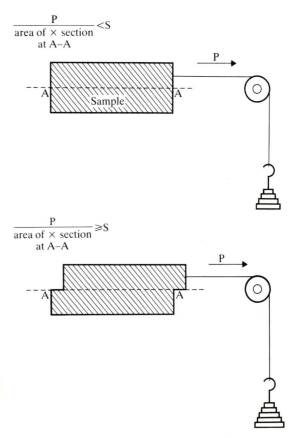

Fig. 4.4 Principle of shear box

While simple, this apparatus is limited in application, since it does not permit accurate reproduction of conditions in the ground, for example, the pore water pressure.

A more sophisticated apparatus, the triaxial apparatus (Bishop and Henkel 1969) is therefore commonly used (Fig. 4.5). The stress to cause sample deformation and failure is applied through the piston. It is not, in this case, immediately obvious that we are concerned with a shear stress and failure 'in shear'. However, when the load becomes great enough the sample will deform or perhaps separate, by shearing along one or more planes. The shear stress is determined by the orientation of the planes in relation to the direction of the applied load.

In the triaxial apparatus, pore water pressures appropriate to those existing in the field can be induced in the sample by varying the pressure on a connecting waterline (Fig. 4.5). Raising the pressure on the oil which normally fills the cell produces an 'all-round' pressure on the sample. The oil is prevented from entering the sample by a rubber membrane fitting the sample closely. The cell is called a 'triaxial' cell since it allows proper control and assessment of the three principal stresses, which describe the 3-dimensional stress state for any element (see e.g. Carson and Kirkby 1972). For some materials a simpler form of test known as the uniaxial compression test is appropriate. There is then no 'all-round' pressure applied. There is also no control of pore water pressure.

Depending on the purpose, the deformation and strength of earth materials can be investigated in two distinct ways. We can apply

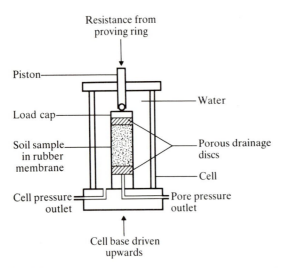

Fig. 4.5 A diagram of triaxial apparatus for measuring soil strength

selected increments of stress and observe the deformation. Alternatively, we can use a machine which inflicts upon the sample a constant rate of strain. In the shear box, this would be a constant rate of travel of one half of the box relative to the other, and in the triaxial test, a constant rate of movement of the piston into the cell. Such a machine (which can be adjusted to give very low strain rates) automatically applies the stress which is needed at any moment to maintain the selected rate of strain. The stress is measured continuously with a device called a proving ring.

4.5 Magnitude and origin of frictional strength

Values of ϕ (cf. e.g. [4.2] or [4.7]) typical for sands are about 30° if loosely packed, and 40° if dense. The same applies for silty soils although these generally have somewhat lower values for the densely packed state than sand. Clays have values from less than 20° to more than 40°.

Frictional strength is in part the resistance to sliding generated where mineral surfaces are sliding over one another (with or without some

water between them). The nature ('roughness') of the mineral surface varies with type of mineral. In soils, so-called interlocking friction, generated as grains attempt to ride over one another, is also important. It is clear that not only mineral composition, but size and shape of the soil grains, and the degree of packing affect ϕ. Not only will ϕ vary from material to material, it is also a somewhat variable property for a particular soil. The process of *initiating* movement results in an over-riding of particle by particle and a change in packing, at least when the effective normal stress is not too high. Consequently the value of ϕ is often to some extent a function of the deformation which the soil has experienced. In this context it is convenient to introduce the precise definition of strain, which is simply the movement, deformation or displacement expressed as a fraction of the original length or volume. Thus linear strain is $\frac{\Delta l}{l}$ and volumetric strain $\frac{\Delta V}{V}$.

Because of the tendency of clays to change volume (p. 52) the range of ϕ for a single clay under various conditions, is greater than for 'non-compressible' soils. A value of ϕ of zero is sometimes described for clay-rich soil but this refers to a particular condition in which the total normal stress is carried by the water. The effective stress to which the material is currently subjected, and also eventual higher effective stresses in the past have significant effects on ϕ. Although clearly a property of the material, ϕ is so dependent on the circumstances and history of the material that further consideration is deferred until Chapter 9. In that chapter the role of frictional strength is analyzed from the point of view of field conditions, and its significance in slope form.

4.6 Cohesive strength

Assessment of strength in clay materials is usually made more difficult by the presence of an additional component, the cohesion. This is not directly dependent on the effective normal stress, as illustrated in Fig. 4.6, where even for $\acute{\sigma} = O$ there is a strength C, due to cohesion.

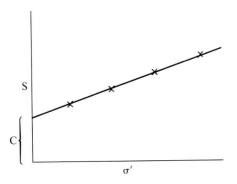

Fig. 4.6 Typical results of test to measure strength of clay as function of effective normal stress

Cohesion is exhibited by any soils which contain significant quantities of fine-grained material (< 2 micron diameter). They are known as cohesive soils, and it is not coincidence that the same soils are generally classed as 'compressible'. Cohesion has its origin in forces arising on mineral surfaces, usually in association with adsorption of water. Consequently cohesion is especially associated with soils of large specific surface area (p. 18), that is, with small particles. Large surface area results in large amounts of adsorbed water, in layers around the particles. Because these encircling 'shells' are not entirely rigid, and also to some extent because of the flexibility of the flaky clay mineral particles, the soils can be compressed, with the consequent expulsion of water. The forces of adsorption and those acting between proximal mineral surfaces are ultimately molecular although not fully understood (see also p. 17). They are to some extent the same as those constituting friction. Nevertheless the fact illustrated in Fig. 4.6 that there is a component of strength, in addition to friction proportional to the normal stress, is very important. The strength of a cohesive soil is given by the classic Coulomb's equation:

$$S = C + \sigma \tan \phi. \qquad [4.9]$$

The magnitude of the cohesion component of strength varies greatly. It can be 300 to 400 10^5 N m^{-2} in some rocks (see examples, Table 9.1). Clay soils often show less than 1 10^5 N m^{-2}, and in some situations considered in Chapter 9, clays have virtually no cohesion. Frictional strength increases with depth according to [4.6] and [4.9], and may well be 1 10^5 N m^{-2} at, say, 10–11 metres depth. Thus, in soils, cohesion tends to be a smaller component of strength than friction, except fairly near the surface.

Changes in composition of a cohesive soil, which in a geological or lithological sense might be regarded as very small (e.g. a change in the concentration of dissolved salts in the pore water) can have highly significant effects on the cohesive strength. Various such changes, which are going on continually under natural conditions in the near surface layers of the earth, assume great importance in analysis of what might otherwise be regarded as 'spontaneous' soil movements due to loss of strength in slopes which have been stable for tens or hundreds of years. Factors influencing cohesion are also further considered in Chapter 9.

4.7 The role of negative pore water pressures

Above the water table, and also under certain other conditions, the porewater pressure has a value less than atmospheric. Associated with the development of such negative porewater pressures is an easily observable increase in strength. Sand on the beach just covered by the lapping sea is soft; as the tide goes out, the sand, some tens of centimetres above the level of the sea water, is quite hard. The water table has fallen somewhat, and the water in the soil above it is held by capillarity (see Ch. 6) and has less than atmospheric pressure. Only when the sand has dried out is it again relatively soft.

So long as the material remains saturated, the effect of the negative water pressure is fully explained by [4.4] or [4.5]. The porewater pressure U having a negative value, the numerical value of U is to be *added* to the total stress to give the effective normal stress $\acute{\sigma}$, and in turn a greater frictional strength component $\acute{\sigma} \tan \phi$. The particles are being pulled together.

At lower moisture contents when air occupies a significant part of the pore space, eqn. 4.5 no longer holds. Although the porewater pressure becomes steadily more negative as moisture content falls, a factor χ (Bishop 1960) must be applied in the effective stress equation:

$$\acute{\sigma} = \sigma - \chi U \qquad [4.10]$$

The value of χ is not easily assessed, but is less than 1. If the pressure of the air in the pores is other than atmospheric this must also be taken into account. Once there are a significant number of pores containing air, the remaining water is filling only the smaller pores or is held on the surface of particles, or in small re-entrants between particles. In any case the remaining water is close to the particle surfaces and likely to be under the influence of adsorptive forces emanating from those surfaces. The effect of the forces is to produce the attraction between water and mineral surfaces which is of the nature of cohesion. The presence of a small quantity of water therefore produces a cohesive strength component. The hardness of the ground during dry summer weather reflects this.

In the past, some authors have associated the occurrence of strength due to water in any state of negative (i.e. less than atmospheric) pressure as being 'cohesive'. Although the increasing moisture content at which the cohesive effect gives way to the frictional effect may be uncertain, there is no reason to regard the existence of moderate negative pore pressures as having any effect on the strength other than through the effective normal stress. The fact that as soon as the pore pressure falls to less than atmospheric it has a negative value, only reflects our rather arbitrary decision to consider atmospheric pressure as zero.

It may be argued that if at low moisture contents the water is instrumental in producing cohesive strength, then that strength component should also be present at high moisture contents since water is then equally present on particle surfaces, in re-entrants between particles, etc. The weakness of this argument is that compressible soils (which are those showing significant cohesion) will normally have a greater volume at higher than at lower moisture content. The distance between particles is greater on average and the surface forces are then less effective, that is, the cohesion is less for the condition of high moisture content.

4.8 Volume change: consolidation

The propensity of many soils to change volume does not arise because of the compressibility of the individual soil components. Both water and soil minerals are essentially incompressible so far as most earth science considerations are concerned. Compaction by engineering procedures, and other natural or man-induced consolidation involves a closer packing of the soil particles. If the soil remains saturated a closer packing must involve the loss or expulsion of water. In engineering practice 'consolidation' normally refers to such a volume change of a saturated cohesive soil occurring under the application of a static load. Tighter packing of dry particles, and also of saturated samples of non-cohesive material can be induced by tapping or vibration, but this process is distinct from consolidation so defined, where the particles do not change their position relative to one another to so large a degree. In geology the term consolidation is used in a different sense to refer to a hardening process involving cementation of particles by weathering and other changes; but this sense will not be considered further here.

Consolidation gives rise to settlements of ground and the effects of the weight of large buildings on clay-rich soils in this connection is of fundamental importance to the foundation engineer. The importance of an understanding of the phenomena of consolidation extends much further however, than the direct implications of foundation settlements.

The consolidation characteristics of a soil are usually measured in a consolidometer, consisting basically of a piston applying load to a compressible soil sample, from the base of which water may drain away through a porous supporting filter (Fig. 4.7). A corresponding volume change occurs in the sample, which is shown by change of height. When the latter has ceased, after 24 hours or so, a further increment of load is added and consolidation continues. The process is repeated until the

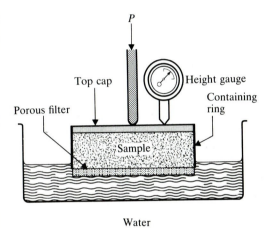

Fig. 4.7 Principle of consolidometer

highest desired pressure is reached. The results are normally plotted as a graph of the logarithm of the applied pressure P, against the void ratio, e. Representing the volume change by the void ratio directs attention to the fact that it is the volume of voids or pores which is changing. The expression 'e log P' curve is widely used, to describe such a graph, but it is not entirely desirable, in that it should be made clear that it is the effective stress $\acute{\sigma}$ that controls consolidation. The pressure applied through the piston constitutes a total stress σ. It happens to equal the effective stress at the completion of each increment of consolidation only because the porewater pressure is then effectively atmospheric (the porewater being continuous with that in the dish and in hydraulic equilibrium with it).

Examples of consolidation curves are shown in Fig. 4.8. A typical curve, A, is shown, with a relatively marked inflexion. The latter is generally interpreted as the effect of an earlier consolidation, produced by effective stresses which reached a value of P_c (= 3 10^5 N m^{-2} in fig. 4.8). This value is often referred to somewhat ambiguously as a *preconsolidation pressure*.

It follows that the consolidation process is not totally reversible. If an effective stress is gradually built up on a typical clay, inducing consolidation, and then removed increment by increment, the clay will tend to expand by taking in water, but along a different path (Fig. 4.8 curve B). This is an example of *hysteresis*, or incomplete reversibility. The void ratio therefore does not depend uniquely on effective stress. It also depends on whether the particular void ratio was reached by *increasing* or *decreasing* effective stress (i.e. loading or unloading), and the effective stress history.

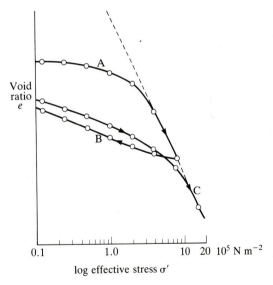

Fig. 4.8 Results of consolidation test. Curve (A) initial test. Curve (B) expansion of material on unloading. Curve (C) consolidation on reloading

If, after removing the effective stress completely (that is, reducing P in the consolidometer to zero), it is once again built up, consolidation occurs in the manner shown by C. Once the previously attained maximum value of $\acute{\sigma}$ is exceeded consolidation continues along an extension of curve A. Subsequently the sample has a new preconsolidation pressure. The steep part of the curve A is called the *virgin consolidation curve*.

Similar volume change behaviour can be produced by a different procedure in which instead of applying a load P, the pressure on the water in the porous base is progressively lowered to negative values, i.e. less than atmospheric. This is what occurs in the suction plate test described in Chapter 6, except that there the compression of the sample takes place laterally as well as vertically. Such tests are carried out to determine the relationship of water suction (that is, negative values for pore pressure) and moisture content, but the water content is measured rather than the volume change. So long as the sample remains saturated the water content changes are, of course, a direct measure of the changes in void ratio. Observations from a suction moisture content test can therefore be converted into a point on the consolidation, or 'void ratio e − log $\acute{\sigma}$', curve (again, provided the sample remains saturated). It is merely necessary to calculate the void ratio corresponding to the moisture content.

It may be a matter of surprise that for many years suction-moisture content relationships were studied intensely by soil physicists in connection with water retention, while engineers were intensely concerned with consolidation tests, without either group realizing that so long as the soil was saturated they were effectively studying the same phenomenon.

The interchangeability of the results is the surest proof that suction, whether produced by drainage or evaporation, produces an equivalent effective stress. In turn the effective stress produces a volume change (consolidation). Clearly also, the strength of the soil is increased. The significance of the entry of air into the soil pores, which can never happen in the consolidation test but which occurs on attaining a certain suction value in the suction-moisture content test, was

considered in relation to the effective stress, in section 4.7.

Under natural conditions, therefore, the effective stresses developed in the soil result from, on the one hand, the weight of soil material overlying the material in question, that is the value of the total stress σ (which is less than $1 \ 10^5 \ N \ m^{-2}$ for near surface layers). On the other hand there is the porewater pressure, which may be highly negative due to the drying or drainage of the soil and giving effective stresses $\acute{\sigma}$ often substantially greater than those arising from overburden.

For the majority of clays, an effective stress of $5 \ 10^5 \ N \ m^{-2}$ (for example) will produce a consolidation up to perhaps 10 per cent of the original volume (that is, of the volume at $\acute{\sigma} = 0$). For one group of clays, the 'swelling clays' which commonly have very high water contents, much greater volume changes occur (p. 81). Indeed the volume may decrease to less than half the original. The characteristic of these clays is an affinity for water arising from osmotic forces associated with cations held on the clay mineral surfaces. In the absence of any effective stress, they will adsorb almost unlimited quantities of water if this is more or less pure, because even large quantities of water do not neutralize the osmotic attraction. Buildings on such materials may settle drastically as the clay consolidates.

Consolidation phenomena are extremely important not only in the practical significance they may have, but also in their effect on other soil properties. The closer packing in a consolidated soil not only directly determines the density and water content of the soil; it also results in increased cohesion because of the greater effect of the particle surface forces in a close-packed condition.

Clearly the degree of consolidation depends on the recent, or more distant, effective stress history of the material which in turn is a product of the situations, the depths of overburden, fluctuations of water table etc. to which the material has been exposed. These in turn are the product of geomorphological and climatological history in the broadest sense.

4.9 Modes of deformation

The nature of the responses to applied stresses constitutes the dynamic properties of a soil. The term 'dynamic' is sometimes applied exclusively to very rapidly fluctuating stress applications, as for example, in earthquakes, or vibrations due to machinery, but these are not considered in this book. The deformations that take place at stresses *less* than those corresponding to the strength as defined earlier (p. 46–7) are however of fundamental importance.

If the applied stresses are sufficiently small there may be no significant deformation (or strain). Reactive stresses are then developed in the material, which are sufficient to hold all particles in place producing a condition of absolute stability. Such a condition is rarer than may be thought. On a historic or geologic time scale most soil materials show *creep* phenomena. These are small displacements, up to a few cm per year, but often immeasurably small except when cumulative over long periods of time. They may occur even though shear stresses in general are far below those required to produce a major movement. Some forms of creep extend to great depths. The effects may be due for example to the potentially enormous stresses associated with thermal expansion or contraction. They may also be associated with chemical change and other long-term effects in addition to those of gravity, effects which are associated with the dynamic nature of the earth's surface region. Climate, cyclical temperature and moisture content changes and ultimately the constant input of solar energy are all causes. Creep can be an important element in erosion, and can also present geotechnical problems (there is a good review paper by Kojan 1967). Quite frequently it is important in causing changes leading to much more obvious, landslide or other movements.

The larger and more sudden displacements of soil that are of immediate concern from a geomorphological or geotechnical point of view, may be successfully investigated experimentally, this involving strength-measuring devices of the kind described earlier. The response of a soil sample to applied stress depends on the nature of the soil itself. Sand soil, at least in small samples, is both soft and friable. Even when damp, sand cannot be shaped by kneading. Clay soil, a cohesive material, can usually be shaped, and exhibits stickiness and adherence.

When movement of soils occurs on slopes producing landslides, mudflows or similar movements, these differences may evidence themselves. In friable material there is a greater chance that movements will be translational, that is showing a definite shear plane, above and below which the material is essentially undisturbed. In soft, 'sticky' material the displacement may take the form of a viscous flow; however, shear planes may also be visible in such material. Exactly which form displacements will take in a field situation depends on various factors effective at the time of failure and not solely on the properties of the material itself. In any case analysis involves determination of shear stresses, strains, and soil strength. For the analysis of the stresses we assume the existence of shear planes, even though there may be no discrete shear plane developed.

At the moment of stress application to a soil sample there is always at least a tiny, usually reversible deformation. The subsequent course of events may involve deformation which attenuates, that is it soon ceases even though the stress is maintained. Alternatively it may be non-attenuating, with deformation continuing at a steady or accelerating rate (even though the shear stresses may decrease). The latter situation, of course, is often of practical significance. On the other hand a rather small amount of deformation may result in a decline of the stress and restoration of stability without adverse consequences. Ultimately, of course all mass movements end in a new balance of stress and resistance.

There is a wide variety of rheological behaviour (rheology being the study of flow and deformation, associated with application of stress to materials), and such is the variety of naturally occurring earth materials that it is important to be aware of the diversity of their rheological properties. Further consideration of our use of the word strength is required. From a geotechnical point of view, constructions are normally designed to allow for some degree of movement, but it is important to know that this amount will not be exceeded. A somewhat special case is that of consolidation, which however illustrates an important point. It constitutes a deformation, but with the 'unusual' feature of a component of the soil, the soil water, leaving the newly-stressed region. In building on clays the engineer allows for consolidation, and maybe for some lateral spreading of the soil mass. The latter constitutes deformation in the more usual sense. From a geotechnical point of view the expected occurrence of consolidation is not regarded as overcoming the strength or bearing capacity, and even less as 'failure'.

Should 'strength' be equated with those shear stresses initiating non-attenuating deformation? Such a point is somewhat academic. Suffice it to say that our definition of strength is quite often anthropomorphic, that is, it relates to our human point of view, rather than to a single well-defined property of the material. What we can conclude is that as scientists, we must be concerned with the nature of the reaction to stresses of the materials we study.

4.10 Methods of describing rheological behaviour

The bending of iron, the flowing of paint, and the slumping of poured dry sugar are examples of diverse rheological behaviour. Such behaviour can be described graphically by

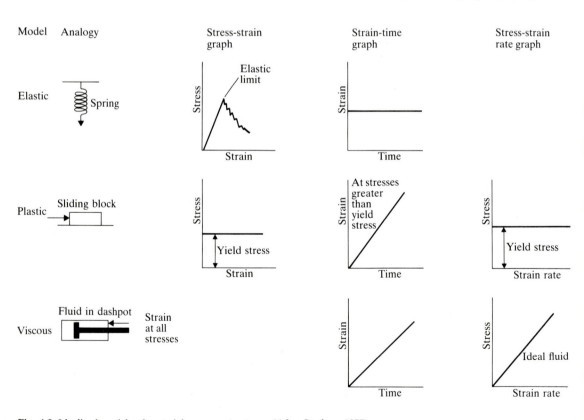

Fig. 4.9 Idealized models of material response to stress. (After Statham 1977)

relating stress to strain (deformation), and these two parameters to time. It can be described by equations. Models or analogies may also be used which involve the well-known behaviour of springs, of dashpots (an arrangement involving a piston which may only be moved against the resistance of a leaking fluid), and sliding blocks (involving friction). In Fig. 4.9, rheological behaviour commonly observed in earth materials is illustrated by the first and last methods of description.

So complex are earth materials and their behaviour that such descriptions are often not entirely accurate; more and more complex combinations are sometimes made to try to explain observed behaviour. Only some of the more important aspects of rheological behaviour can be reviewed here, but the reader is referred to the chapter in Statham's book (Statham 1977) – from which Fig. 4.9 is adapted.

Ideal elastic behaviour is shown in a very minor degree by most soils or rocks but the instantaneous deformations (where strain does not increase with time, Fig. 4.9a and which are reversible) are normally so small as to be insignificant. They are often masked by other deformations occurring subsequently. Viscous behaviour, which occurs in mudflows, is characterized by flow continuing as long as the stress is maintained, with the rate of flow (that is, rate of strain) increasing as the stress increases. In ideal viscous behaviour, very slow flow occurs with extremely small stresses. A well-known example is asphalt which although appearing hard, slowly deforms under its own weight. Viscous flow is associated with fluids and the relatively slow flow of asphalt compared to say, water, simply reflects the higher viscosity of the former. Ice and frozen ground show substantial viscous behaviour, and certain forms of slow creep of soil masses on hill slopes of very slight angle (and therefore with low shear stresses) can also be so regarded.

Purely plastic behaviour is characterized by continuous deformation but only at stresses greater than a certain threshold value, the yield

stress, below which there is no deformation. Amorphous materials of a colloquially 'plastic' nature deform plastically if they do so in the manner described. Contrary to everyday notions of 'plasticity', the sliding of a rigid block of rock or soil once initiated, is an example of 'rigid plastic' behaviour.

Some materials actually produce greater resistance *after* some deformation has occurred and are referred to as 'strain hardening'. More commonly, earth materials show a loss of strength as strain proceeds: they are 'strain-softening'. The latter phenomenon has importance in relation to the stresses required to produce 'failure', that is, to overcome strength in the sense of producing a large, rapid and continuing strain or displacement and attendant loss of resistance.

Fig. 4.10 is an example of a stress-strain relationship which occurs frequently in cohesive soils. The behaviour is in fact, important in providing a rational explanation for the particular nature of landslides in many clay

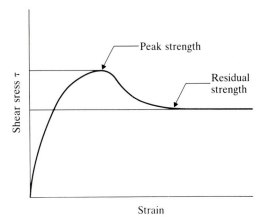

Fig. 4.10 Typical behaviour of clay rich soil, in strength test with constant rate of strain

soils. The sample is strained at a constant rate. The stress, necessary to maintain the strain rate, rises rapidly initially. The stress continues to rise, subsequently, although the increment of stress per unit of additional strain decreases. A point is then reached at which strain continues without additional stress, and indeed the stress decreases somewhat. The highest stress reached is the *peak strength*, while the lowest stress subsequently reached is called the *residual strength*. Sometimes there is little or no difference between the two. In some cases the residual strength is a small fraction of the peak strength. The difference is mainly due to the absence of cohesive strength in the residual strength value. The values are to some extent dependent on the rate of straining, or the rate of loading if this is the controlled variable in the test.

This brief discussion has emphasized the complex nature of the responses to stress exhibited by earth materials. They rarely exhibit what may be regarded as ideal viscous or plastic behaviour, nor even a simple combination of such behaviours. It is therefore essential to be aware of the various possibilities for deformation behaviour, and how these relate to different earth materials, and to the various patterns of stress application occurring under natural conditions. In Chapter 9 various natural situations are considered, in order to illustrate the manner in which mass movements occur, their importance in relation to stability, and the formation of slopes.

5

Thermal properties

5.1 Significance of soil thermal properties

We associate the word 'thermal' with heat, energy, warmth and temperature. Much uncertainty surrounds our use of these and similar words. A farmer may describe part of his ground as being 'cold'. Usually he refers to a spot where spring germination and growth is delayed. Germination and growth periods commence at certain temperatures which constitute threshold values. Nevertheless, the soil that is 'cold' to seeds may have the same mean annual temperature as other neighbouring ground of different lithology. It may also show a lesser depth of penetration of frost in the winter and, at least just below the frozen layer, it is then warmer than neighbouring soils. At any time there are temperature gradients; the ground does not have a single temperature. The distribution of temperature relative to depth depends on soil type, that is, the soil's thermal properties. At the same depth, with the same conditions of surface cover and climate, different soils will show different temperatures.

Another uncertainty concerns the significance of temperature: a bright sunny day with an air temperature of -1 °C is pleasant, but the same temperature in a blinding snow storm constitutes a decidedly 'cold' day. Just as we are concerned with exchange of heat from our bodies to the surroundings, in the same way below the ground surface, it is heat transfers which are primarily of importance. Indeed, the temperature measured at a point in the ground is dependent on the heat that has been transferred to or from the point. While heat flows as a result of a temperature gradient, and the amount increases proportionally to the gradient, the rate of transfer depends on the thermal property of the material known as *thermal conductivity*.

Consider a piece of cork, or asbestos, or foam plastic. If a hot body is placed in contact with one side, relatively little effect is felt at the opposite side. If the material is replaced with a similar sized body for example, steel, then a much more marked temperature effect is rapidly felt. Cork, asbestos, etc. have low thermal conductivity, while steel has relatively high thermal conductivity and heat is rapidly transmitted through it.

When heat is transferred to or from a substance, the temperature changes by an amount which depends on the nature of the substance. This property of the material constitutes its *heat capacity*. If equal quantities of oil, and of water, in similar containers are warmed on identical hot plates, the temperature of the oil rises about twice as fast as that of the water. The heat capacity of water is almost twice that of oil. The natural variability of earth materials is such as to give a wide range of thermal conductivities and heat capacities.

Ultimately temperature and heat storage fluctuations in natural ground are due to weather and climate. Fluxes of heat between the atmosphere and ground are continually controlled by exchange processes in the surface region, which vary with different vegetation or other surface features, and with soil type. The distribution of heat and temperature *within* the ground are essentially controlled by the soil thermal properties of conductivity and heat capacity.

It is of practical importance that we understand ground thermal regimes. The engineer designing a buried high voltage transmission line is concerned that resistance heating might lead to damage of the cable. Whether the heat liberated is satisfactorily dissipated, depends on the thermal conductivity of the soil, that is, its ability to transmit heat away at a satisfactory rate, and its heat capacity, that is, on its ability to absorb heat without undue temperature rise. The design of heating or cooling systems for underground structures similarly involves the temperature and thermal properties of the surrounding ground. Systems for utilizing solar energy, whether for domestic or other purposes, commonly involve heat storage facilities in the ground. Conditions for biological processes, as

well as weathering and other natural phenomena depend on the thermal regimes.

Although the distribution of temperature with time and depth is an important parameter, scattered temperature observations without knowledge of the soil thermal properties are rarely meaningful. A continuous measurement of temperatures through a period of a year or more may permit calculation of thermal properties in that the temperatures are a function of these properties. It is usually the case that direct determination or estimation of thermal properties are necessary to a proper understanding of the *ground climate*.

5.2 Thermal conductivity

Heat is a form of energy, measured in joules, and is said to flow – from higher to lower temperatures. Consider a uniform solid bar of some material, one end of which is maintained at a certain temperature, and the other at some lower temperature (Fig. 5.1). The flow of heat along the bar depends on the difference between the two end temperatures, T_1, T_2; the distance between them (the length of the bar) L; the cross-sectional area, A, of the bar (a thicker bar could clearly transmit more heat); and on the properties of the material composing the bar, as characterized by the *thermal conductivity coefficient* λ. The flow of

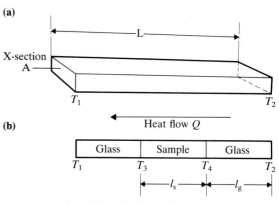

Fig. 5.1 (a) Dimensions relevant to the flow of heat in a uniform bar. Temperature T_1 is less than T_2
(b) Principle of divided bar apparatus, for determining thermal conductivity

heat will clearly also depend on the interval of time involved. Thus the flow of heat is given by:

$$Q = A \lambda \frac{T_1 - T_2}{L} \qquad [5.1]$$

with units Q, joules s^{-1}
A, m^2
$T_1, T_2,$ °C
L, m
λ, J s^{-1} m^{-2} °C^{-1} m
$= $ W m^{-1} °C^{-1}

The values of thermal conductivity for a number of common materials are given in Table 5.1.

A comparable experimental arrangement known as the divided bar or plate apparatus is used for determination of λ in soils and rocks (Birch 1950). In its simplest form the apparatus involves placing the sample between plates of glass or other uniform solid (Fig. 5.1b) whose conductivity is accurately known. Constant temperatures T_1 and T_2 are established and after a short while a steady state condition is attained. The flow of heat q_s though the sample is then necessarily the same as through the glass plates q_g. The temperatures T_3, T_4, at the faces of the sample will assume values so that:

$$q_g = A \lambda_g \frac{T_1 - T_3}{\ell_g} = A \lambda_s \frac{T_3 - T_4}{\ell_s} = q_s \qquad [5.2]$$

If the thickness of the sample, ℓ_s, and glass plate, ℓ_g, are equal then,

$$\frac{\lambda_s}{\lambda_g} = \frac{T_1 - T_3}{T_3 - T_4}$$

An alternative device (Jackson and Taylor 1965) used with soils is the line heat source probe which is inserted into a soil sample. The probe is basically a length of steel tubing inside which is a heating coil, and a temperature measuring device. A known quantity of heat is supplied by a monitored electric current through the coil. The rate of dissipation of the heat depends on the conductivity of the soil, and is shown by the rate of temperature rise of the probe.

There are several other methods including

Table 5.1 Thermal conductivities for some common materials

	Conductivity W m^{-1} °C^{-1}	Source
Concrete, ballast 1 : 2 : 4	1.5	Kaye and Laby (1973)
Concrete, cellular	0.1–0.2	Kaye and Laby (1973)
Cork (baked slab)	0.038–0.046	Kaye and Laby (1973)
Glass, double extra dense flint	0.55	Kaye and Laby (1973)
Glass, Pyrex	1.1	Kaye and Laby (1973)
Paraffin wax	0.25	Kaye and Laby (1973)
Polystyrene, cellular	0.035	Kaye and Laby (1973)
Iron	83.5	Kaye and Laby (1973)
Ice	2.3	Kaye and Laby (1973)
Water	0.56	Kaye and Laby (1973)
Wood	0.3	Touloukhian *et al.* (1970)
Minerals		Averages of typical values, collected by Johansen, 1975. In some cases reported values (at temp. $\approx$ 25 °C) are greater or lower by 25 per cent
Pyroxene	4.4	
Amphibole	3.5	
Chlorite	5	
Felspar	2.0	
Quartz	8.0	
Clay minerals (unspecified)	2.9	
Mica	3	(reported values [Touloukhian *et al.*, (1970)] range from 0.04 to 40 – depending possibly on entrapped air).

(Many additional minerals and values are given in Horai (1971).

Examples of rocks		
Granite	2.0	Touloukhian *et al.*, 1970
Limestone	1.0	Touloukhian *et al.*, 1970
Gabbro	2.0	Touloukhian *et al.*, 1970
Limestone	2.9	Judge, 1973
Quartzite	5.9	Judge, 1973
Shale	1.5	Judge, 1973

Thermal conductivity varies with temperature and pressure and sometimes with orientation of mineral axis etc., but the variations are not usually significant in studies of earth materials. For many earth materials, however, there is a large variation in values reported, even for similar conditions. The values given are to be regarded therefore as approximate (some directly quoted figures are not rounded off). References contain additional data.

one involving determination of the diffusivity of the soil (see section 5.7), from which the conductivity is easily obtained if the heat capacity of the soil is known.

5.3 Conduction of heat in soils and rocks

Most soil or rock materials present considerable difficulties for determination of conductivity. These arise particularly because of the composite nature of the materials. Conduction of heat occurs at different rates through the water, air and soil mineral constituents and this is considered in theoretical methods for estimating thermal conductivities. In addition the composition of a soil or rock is rarely constant. Similar materials will have different water contents at different times of the year or in different locations.

This limits seriously the use of experimental methods of determination. A single determination of λ for a soil has limited value. It is necessary to have some understanding of the effects of water content, mineralogy, the degree of packing (consolidation) etc. before reasonable estimates can be made of the value of λ appropriate for specific situations.

Heat flow in porous material is not limited to 'simple conduction' as in more homogeneous materials, but also occurs in association with movement of water, vapour or air, these carrying heat with them. This will obviously be the case where for example, infiltration is occurring. But more important is the fact that a temperature gradient itself initiates movements of vapour (and to a lesser extent of water) initially at equilibrium. In soils that are not saturated, a sequence of vapourization (involving adsorption of latent heat of vaporisation), vapour flow, and then

condensation (involving release of latent heat) occur in air and vapour-filled pores, and the exchanges of latent heat are superimposed on the heat conduction in the strict sense.

Nevertheless, theroretical investigations coupled with results of experimental studies, have provided a basis for estimation of thermal conductivity coefficients, suitable for many practical purposes. The knowledge that the coefficients may be affected by processes other than simple conduction helps to avoid gross errors in such estimations.

5.4 Assessment of thermal conductivities based on soil composition

Various authors have devised models for the conduction of heat in granular materials with a view to predicting thermal conductivity on the basis of the composition of the material (De Vries 1952, Woodside and Messmer 1961).

The models are not totally reliable, large errors occurring with certain kinds of soils. Saturated soils are the most amenable to prediction of thermal conductivity because heat transfer occurs mainly by simple conduction. The absence of air or vapour-filled spaces obviates vapour transfer and latent heat effects. The application of temperature gradients to saturated soils does not usually cause important movements of water.

A comprehensive study by Johansen (1972, 1973, 1975) led to the conclusion that the conductivity of saturated materials is fairly well represented by the geometric mean equation:

$$\lambda = \lambda_1^{V_1} \ \lambda_2^{V_2} \ \lambda_3^{V_3} \dots . \qquad [5.3]$$

where λ is the conductivity of the soil, and λ_1, λ_2, λ_3, etc. the conductivities of the constituent soil minerals and water. V_1, V_2 etc., are the volume fractions (i.e., the volume per unit volume of soil) of each of the constituents. The equation gives large errors when there is a large difference between the lowest and highest conductivity; hence, the air present, with its very low conductivity, makes the equation unsuitable for unsaturated soils.

Johansen used the extensive experimental measurements of Kersten (1949) to verify a series of relationships between various factors and soil thermal conductivity. He found that the mineral composition of soil, especially the amount of quartz, is of particular importance. Quartz has a relatively high conductivity, about 8 W m^{-1} °C^{-1}. Other common soil minerals, notably feldspars and micas, have $\lambda \simeq$ 2 W m^{-1} °C^{-1} (Table 5.1). Thus it is often convenient to consider the soil mineral component as consisting of only two volume fractions, the quartz fraction, and the remaining mineral(s) fraction. Applying [5.3] to saturated soils gives results for the conductivity accurate to perhaps ±10 per cent. Knowledge of dry bulk density ρ_d, and water content is required in order to obtain the volume fractions of each component.

The conductivity of completely dry soils is approximated, with a lesser degree of accuracy, by an empirical equation:

$$\lambda = \lambda_a \, n^{-2 \cdot 5} \qquad [5.4]$$

where λ_a = conductivity of air 0.022 W m^{-1} °C^{-1} at 0°C
n = porosity, $1 - \rho_d$
ρ_s = density of mineral component

The conductivity of a soil in a damp or wet but not saturated state, lies between the extremes of dry and saturated and increases with the degree of saturation S_r (with the qualification for saturated swelling soils noted below). This follows from the much higher thermal conductivity of water compared to air. When the moisture content is low and vapour and air-filled passages are continuous, some of the heat transfer is by processes of vapour diffusion, evaporation and condensation. The 'conductivity' coefficient is then to some extent dependent upon temperature gradient. Consequently discretion is necessary in assessing appropriate values and applying these to field situations.

The importance of porosity n, is two-fold. Firstly the more porous a soil, the greater is the volume of air or water in unit volume of the soil. Both these substances have, of course, lower conductivities than soil minerals. Secondly, the conductivity of a soil also decreases as the porosity increases because the amount of contact between particles decreases. The extent of interparticle contacts, and their nature, are important in respect to heat conduction paths through the particles. In the examples illustrated in Fig. 5.2, curves 2, 4, 6

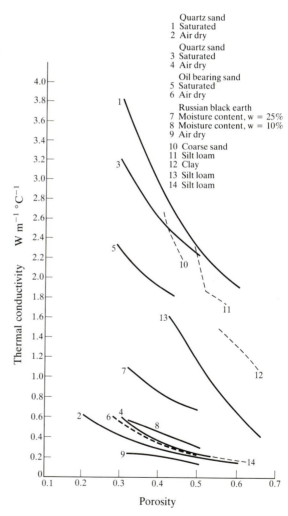

Quartz sand
1 Saturated
2 Air dry
Quartz sand
3 Saturated
4 Air dry
Oil bearing sand
5 Saturated
6 Air dry
Russian black earth
7 Moisture content, w = 25%
8 Moisture content, w = 10%
9 Air dry
10 Coarse sand
11 Silt loam
12 Clay
13 Silt loam
14 Silt loam

Fig. 5.2 Variation of thermal conductivity with porosity, n, for different types and conditions of soil. (After Wood, 1979, from various sources)

and 9 show the effect of degree of packing alone; the remainder are for soils with a moisture content which may have changed as well (information compiled by Wood 1979). While the increasing porosity tends to decrease the conductivity, an increase in water content counteracts the effect to some degree.

Many soils swell in some degree, in association with moisture content changes. When the moisture content of a saturated soil of this type is increased, the conductivity must always decrease. The effect may be large in the case of the true 'swelling clays' (p. 81) in which very high moisture contents can occur. The additional water is not replacing air, but displacing mineral particles (which have higher conductivity).

Johansen developed nomograms (Fig. 5.3) for the assessment of thermal conductivities; the nomograms require determination of the soil's gravimetric properties and of the percentage of quartz mineral content. Although assessment of thermal conductivities in this way may involve substantial inaccuracies, perhaps

±25 per cent with damp (not saturated) soils, the proceduce appears the best available for many practical purposes. Johanson's work has added value in enabling us to visualize how the varying nature of the different soils, affects conductivity.

The following general conclusions are prompted by Johansen's work and seem in accordance with a fairly large number of experimental observations in various laboratories. Some exceptions to these guidelines can be expected. For *saturated* soils:

1. The lowest conductivities are about 1 W m^{-1} °C^{-1}. Such saturated soils are composed of minerals other than quartz and are of low dry density (e.g. $\rho_d = 1.2 \times 10^3$ kg m^{-3}). For similar soils in a highly dense state ($\rho_d = 2.0 \times 10^3$ kg m^{-3}) the conductivity is about 1.5 W m^{-1} °C^{-1}.
2. For saturated soils composed exclusively of quartz grains the conductivity ranges from about 2 W m^{-1} °C^{-1} to about 4 W m^{-1} °C^{-1}, increasing roughly linearly with dry

Thermal conductivity of unfrozen soils

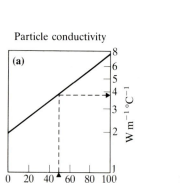
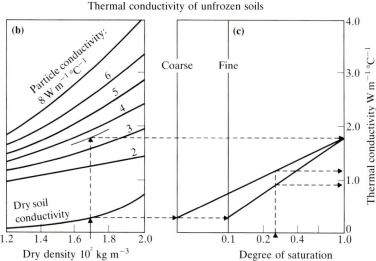

Fig. 5.3 Nomograms for determining conductivity of unfrozen soils. From the percentage quartz content a 'particle conductivity' is obtained (a). From a value of dry density the intersection with the appropriate curve in (b) is obtained. The ordinate value is extended, to intersect the ordinate axis of (c). The value so obtained is the conductivity for 100 per cent saturation. For the soil when essentially dry, extension from the 'dry conductivity' curve in (b), gives the conductivity for the dry state in (c). A line is then drawn connecting the values for 'dry' and 'saturated' allowing determination of conductivity for intermediate degrees of saturation. A distinction is made between coarse and fine-grained soils. The latter (minimum 10 % < 0.06 mm) are assumed in the 'dry' state to have a degree of saturation of 0.1 (After Johansen 1973).

density. This is partly because water has a much lower conductivity (0.56 W m^{-1} °C^{-1}) than quartz and the water content decreases as dry density increases.

3. For soils partly quartz and partly other minerals, the value lies between 1 (case 1 above) and somewhat less than 4 W m^{-1} °C^{-1} (compare case 2). As the quartz content (per cent of mineral content) is increased, the conductivity increases very roughly linearly, although more rapidly for higher quartz percentages. The effect of increasing dry density is a roughly linear increase and for soils with high quartz content the conductivity doubles over a range of ρ_d 1.3 to 2.0 10^3 kg m^{-3}. The effect is less marked for soils of low quartz content.

For *dry* and *damp* soils:

1. Completely dry soils have conductivities less than 1 W m^{-1} °C^{-1}. Dry density and mineral composition affect the conductivity, which may be as low as 0.1 W m^{-1} °C^{-1}.
2. A damp soil has a conductivity between the 'dry' and 'saturated' value for the same material, depending on the degree of saturation. At low degrees of saturation conductivity increases rapidly with moisture content, but for soils approaching saturation variations in water content have relatively small effect.

Table 5.2a Mass and volumetric heat capacities of some common materials After Kaye and Laby (1973)

	c, $J\ kg^{-1}\,°C^{-1}$	C, $10^6\ J\ m^{-3}\,°C^{-1}$ (calculated from values given left)
Brass (red)	377	3.20
Steel, carbon	480	3.74
Glass, flint	500	1.7 (density varies)
Glass, pyrex	780	1.7
Paraffin wax	2900	2.61
Air (dry, atmosph. pressure, 20 °C)	1006	0.000857
Ice	2000–2100	1.83
Water (20 °C)	4181.6	4.18
Water (0 °C)	4217.4	
Sea water	3930	3.93

Table 5.2b Mass and volumetric heat capacities of rocks and minerals

	c, $J\ kg^{-1}\,°C^{-1}$	C, $10^6\ J\ m^{-3}\,°C^{-1}$ (mostly calculated from values given, left)
Augite	810	2.7
Hornblende	820	2.6
Mica	860	2.4
Orthoclase feldspar	790	2.0
Quartz	790	2.1
Basalt	840	2.4
Clay minerals	900 *	
A granite	800	2.1
A limestone	910	2.5
A sandstone	920	2.2
A shale	710 †	1.6

After Forsythe (1969) except:
* Monteith (after Van Wijk 1963).
† From *Handbook Physics and Chemistry*, 1976.
See text for comments concerning accuracy.

5.5 Heat capacity and soil composition

Volumetric heat capacity, C, is the quantity of heat (joules) which raises or lowers the temperature of 1 m^3 of material by 1 °C. Table 5.2 shows volumetric heat capacities of various materials. If the quantity of heat refers to 1 kg the quantity is the mass heat capacity (or specific heat capacity – other terminology may also be met). Generally the volumetric heat capacity is the desired quantity for ground thermal studies.

In heterogeneous materials such as soil the heat capacity is simply the sum of the heat capacities of the proportional parts, thus:

$$C = V_s C_s + V_w C_w \qquad [5.5]$$

where V_s, V_w are volume fractions of solids and water respectively, C_s and C_w are the volumetric heat capacities. The heat capacity of air may be ignored.

In practice a convenient equivalent equation is used:

$$C = \rho_d \left(c_s + \frac{c_w W}{100} \right) \qquad [5.6]$$

where ρ_d is the dry bulk density of the soil kg m^{-3}; W is the water content per cent dry weight; while c_s is the mass heat capacity of the soil's mineral component. The value of c_s for most soil or rock minerals generally lies in a relatively narrow range from 8 10^2 J kg^{-1} °C^{-1} (quartz) to 9 10^2 J kg^{-1} °C^{-1} (clay minerals). The mass heat capacity of water c_w is relatively high, 4.18 10^3 J kg^{-1} °C^{-1}. Organic matter has a fairly high mass heat capacity, about 2 10^3 J kg^{-1} °C^{-1}, but care is needed in definition. The value refers to dried material, that is exclusive of the heat capacity associated with the large amounts of water often loosely held in plant remains, wood etc. This water content will of course be included as part of the water content of the soil.

Heat capacities of a number of common materials, including rocks and rock minerals are given in Table 5.2a and b. Care must be taken not to confuse volumetric heat capacities of rocks or minerals, with the volumetric heat capacity of the same material in the form of a soil. In the latter case, the volumetric capacity is reduced in the dry state by a factor due to porosity, of

$$\frac{\rho_d}{\rho_m}$$

where ρ_m is the specific gravity of the rock or rock mineral. Heat capacities vary with temperature, but not usually significantly over the range of commonly occurring temperatures. The variability within a single rock group is greater. The values given for rocks (Table 5.2b) are typical rather than definitive, and precise values for a particular rock require calculation based on knowledge of the component minerals and porosity. There are also some variations in reported values for minerals, and the volumetric heat capacity is affected by density variations. Values for additional minerals are given in the source references.

In many field conditions, where soils are damp or saturated the heat capacity of the water content is the largest component of the

heat capacity of a soil. The soil heat capacity is greatly affected in the near surface soil layers by seasonal, climatic, or weather-induced variations in moisture content. It is generally not meaningful to speak of the heat capacity of soils unless the moisture content is defined.

It is also generally unsatisfactory to make comparisons of heat capacity of different soils, on the basis of knowledge of moisture content alone. Thus a clay and a sand each having 15 per cent dry weight water content are unlikely to have closely similar heat capacities, and this may also be true for samples of superficially similar materials. The dry densities may be significantly different (cf. Table 4.1) and the heat capacities will differ accordingly ([5.6]).

The volumetric heat capacity of a soil cannot reach 4.18 10^6 J m^{-3} °C^{-1} as this would imply a body of water alone (all other soil components have lower heat capacities). Nor can the value fall below about 1 10^6 J m^{-3} °C^{-1} this implying a dry material very loosely packed. Equation [5.6] allows simple and accurate assessment for most materials.

Where the volumetric heat capacity is required for an incompressible soil over a range of water contents, a determination of dry density ρ_d at one water content is normally sufficient in conjunction with the use of [5.6]. Compressible soils change volume (and thus dry density), depending on the water content, especially when saturated. Equation [5.6] may then be modified accordingly:

$$C = \rho_{d_w} \frac{1}{1 \pm \left(\frac{\rho_d}{\rho_w} \, _w \, \frac{\Delta W}{100} \right)} \left(c_s + \frac{c_w}{100} \, W \right) \quad [5.7]$$

where ρ_{d_w} is the dry density at a certain water content and ΔW is the difference between this water content and the water content of interest, W. The equation is true only so long as the soil is saturated. The term $\frac{\rho_{d_w}}{\rho_w} \frac{\Delta W}{100}$ is positive when

ΔW represents an increase in water content.

5.6 Heat capacity and latent heat, apparent heat capacities

Normally the addition or removal of heat results in a change of temperature, of a magnitude which is determined by the heat capacity of the material. At certain temperatures instead of a change of temperature occurring, addition or removal of heat results in a change of phase. The quantities of heat associated with a change of phase are referred to as latent heat. The latent heat of fusion water to ice is 333×10^3 J kg^{-1}, while to evaporate water requires approximately 2.45×10^6 J kg^{-1}. Changes of phase occur frequently under natural conditions and constitute an important factor in many geomorphological processes. The effect of supplying or removing heat on the temperature cannot be fully treated by consideration of heat capacity alone, if a change of phase occurs, and the quantities of latent heat are often not easily predictable.

Because the volume of water in the vapour phase is so much greater than that in the liquid phase, evaporation or condensation is normally associated with movement of water vapour to or from the body of soil under consideration. Thus to analyze such effects completely consideration would have to be given to the processes of vapour movement in the soil.

The case of freezing is somewhat different. Although freezing or thawing is also often associated with movements of moisture in the soil (see Chapter 7), nevertheless it is useful to consider the effects of freezing temperatures on a soil whose moisture content (water and eventual ice) remains constant. The latent heat released on freezing or absorbed on thawing can then be regarded as a component of the heat capacity. If we consider the steady removal of heat and the effect upon the temperature of a soil sample which is being slowly cooled to 0 °C, we are initially concerned with the soil's heat capacity as unfrozen material. When the temperature reaches 0 °C, or just below, ice starts to form, and the rate of temperature change is much reduced. Much of the heat removed is then associated with the changing of the water into ice. If we did not know what was happening, we would believe the heat capacity had suddenly increased very greatly. In many practical applications it is convenient to so consider the effects and to refer to the apparent, or effective, heat capacity, it being understood that *apparent heat capacity* consists not only of the true heat capacity of the water, eventual ice, and mineral constitutents, but also of that component of heat which is liberated in association with the transformation of water into ice.

In fact the freezing of water in soils does not occur at a single temperature as is the case with pure water at 0 °C. As ice forms progressively in the soil pores the conditions for freezing change and the freezing point also changes. Consequently latent heat of fusion is liberated over a range of temperature of (in many soils) at least several degrees below 0 °C. Apparent heat capacity is very high in the region of 0 °C and falls steadily with temperature until ultimately there is no further transition of water to ice, when we again have the true heat capacity whose value depends on the proportion of ice, mineral soils, and perhaps permanently unfrozen water. Methods of determining the apparent heat capacity are considered in section 7.10.

5.7 Thermal diffusivity

If we consider a body of material initially at a uniform temperature, to the surface of which a temperature change is applied, there will be a flow of heat to or from the surface. Change of temperature within the body occurring as a result of the applied surface temperature will clearly depend on the ease with which heat flows in the material, that is the thermal conductivity. It also depends upon the temperature change associated with the loss or gain of heat by that material, that is the heat capacity. This dual dependence is reflected in the quantity known as the thermal diffusivity, which is the ratio of the thermal conductivity to the volumetric heat capacity, $\frac{\lambda}{C}$, with the units $m^2\ s^{-1}$. This quantity is much used in analysing changes of temperature and heat distribution. In some European languages it is known as the 'temperature conduction coefficient' which emphasizes its importance in relation to temperature distribution. The term thermal diffusivity however emphasizes the analogies between this quantity relating to the distribution of heat energy, and the diffusion coefficient used, for example, in considering the distribution of a solute in a solution as a result of molecular motion. Thermal diffusivities of naturally-occurring soils commonly have values in the range 1 to 15 10^{-7} $m^2\ s^{-1}$.

The significance of thermal diffusivity is illustrated by the Fourier heat transfer equation, applied to the rudimentary situation of changing temperature in a layer of soil. Consider a uniform layer of soil thickness Z, through which heat is flowing. The condition is necessarily not a steady state; the temperature gradient changes with depth and time. Assume the layer is horizontal and that the heat flow is in an upward direction. The heat flow into the lower surface of the layer is then given by:

$$Q_1 = \lambda\left(\frac{\Delta T_1}{\Delta Z_1}\right) \qquad [5.8]$$

and the heat flowing out of the upper surface is:

$$Q_2 = \lambda\left(\frac{\Delta T_2}{\Delta Z_2}\right) \qquad [5.9]$$

where ΔZ_1 and ΔZ_2 refer to small increments at the upper and lower surfaces respectively of Z; and ΔT_1, and ΔT_2 are the temperature differences through these increments (Fig. 5.4). The difference, $Q_1 - Q_2$, is the heat energy being gained or lost by the layer Z in unit time.

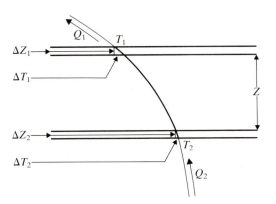

Fig. 5.4 Diagram showing quantities relevant to analysis of changing temperature gradient in layer of thickness Z (see text)

This quantity is also given as the product of the change of temperature ΔT of the layer in time ΔT, its heat capacity C, and thickness Z:

$$Q_1 - Q_2 = -C\frac{\Delta T}{\Delta t}Z = \lambda\left(\frac{\Delta T_1}{\Delta Z_1}\right) - \lambda\left(\frac{\Delta T_2}{\Delta Z_2}\right) \qquad [5.10]$$

in differential form we obtain:

$$\frac{\partial T}{\partial t} = \frac{\lambda}{C}\frac{\partial^2 T}{\partial Z^2} \qquad [5.11]$$

This equation tells us that the rate of change of temperature with time is equal to the diffusivity times the rate of change of the temperature gradient with depth. Thus if the temperature gradient in uniform ground varies with depth (as it usually does, see Fig. 10.2a, Chapter 10) temperatures are changing at a rate dependent on the diffusivity.

In Chapter 10 a few relatively simple equations involving the thermal diffusivity are given, which may be used to describe the effects of cyclic temperature changes on bodies. Ground thermal studies are commonly concerned with diurnal or annual temperature change, and with other cyclic, climatic or weather changes. Distribution of temperature or heat flows in the ground with time and depth are fairly easily analyzed in a general way with these equations, which provide a logical framework for understanding the climate in the ground – the subject of Chapter 10.

A diffusivity of a material can also be *measured* utilizing equations of the same type. If a cyclic temperature change is applied to the surface of a body, and the temperatures at different times and depths are measured, then the diffusivity can be calculated. This being the case it might be thought that most ground thermal problems could be satisfactorily handled by consideration of diffusivity values, rather than of the two properties of conductivity and heat capacity. Particularly because of the variable nature of soils, their moisture content, mineral composition and other characteristics, consideration of the diffusivity alone constricts our understanding of the ground thermal regime and the factors controlling it. It is for example difficult, if not impossible, to visualize directly the effects of porosity, moisture content, or mineral composition on diffusivity. Instead one proceeds to consider the effect of moisture content and other factors on conductivity, and heat capacity, in the manner described earlier, and then on the diffusivity as the ratio of these two properties. Figure 5.5 shows the effect of

moisture content on the thermal diffusivity of various soils. At low moisture contents $\frac{\lambda}{C}$ increases rapidly with moisture content because of the rapid rise in conductivity. At high moisture contents additional moisture increases heat capacity, C, while λ increases little and may decrease. The confusion of curves on this graph, illustrates how difficult it is to lay down general rules for diffusivity in relation to soil type, and moisture content.

5.8 Conductive capacity

The product of the thermal conductivity and heat capacity is of interest in that the total exchange of heat to or from a layer affected by temperature change at its surface is proportional to $\sqrt{\lambda C}$, the *conductive* capacity. Consider, for example, the annual temperature cycle occurring at the surface of the ground. During the summer heat is transmitted downwards from the surface and a corresponding quantity of heat will leave the ground during the winter. The magnitude of the quantity is proportional to $\sqrt{\lambda C}$. Values of the conductive capacity are given in Table 5.3.

Table 5.3 Conductive capacities, $\sqrt{\lambda C}$, 10^2 J m^{-2} °C^{-1} s$^{1/2}$

Wet clay	96–209
Wet sand	50–250
Peat	79–163
Frozen peat	108–222
Ice	217–255
Snow (density 0.2 10^3 kg m^{-3})	8–25

Collected by Williams, G.P. (1970), from various sources

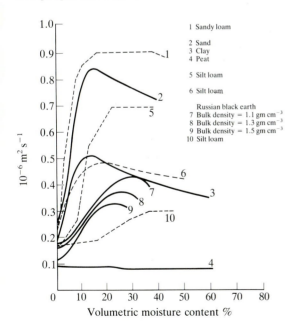

Fig. 5.5 Effect of moisture content on the thermal diffusivity of different soils. (After Wood 1979, from various sources)

The role of the conductive capacity is illustrated by considering the application of a certain increment of temperature to the ground surface. Suppose the soil has a high conductivity and a high heat capacity (and thus a high value of $\sqrt{\lambda C}$). Clearly flow of heat into the ground is aided by the high thermal conductivity. As the heat penetrates the ground, the rise of temperature below the surface will depend on the heat capacity and, because this is high, the rise will be relatively slow. A steep temperature gradient will therefore be maintained in the surface layer of the ground, also facilitating the heat flow. If on the other hand, the soil has a low conductivity the entry of heat is retarded, and if in addition it has a low heat capacity, the surface layer will on this account show a relatively rapid temperature rise. The temperature gradient in the layer will decrease and correspondingly, the heat flow is further reduced – this being the case for a low value of $\sqrt{\lambda C}$. Thus the conductive capacity coefficient relates directly to the ability of the ground to absorb and release heat. Such considerations are important, of course, in the design of solar (or other) heat storage systems in the ground.

The term has also been called the 'contact coefficient', 'thermal property', 'thermal inertia' etc. (Businger and Buettner 1960). Although a function of material properties it also relates particularly to temperatures generated at boundaries or interfaces, for example between air and ground, as a consequence of fluxes within the two media. The rate of temperature change at a boundary is inversely proportional to $\sqrt{\lambda C}$. These effects are well illustrated by Businger and Buettner (1960) who suggest comparing the effect of touching bodies of wood, copper, asbestos, etc. each uniformly raised to 50 °C. Whether the surface of one's finger becomes unpleasantly hot clearly depends on the nature of the body, that is, $\sqrt{\lambda C}$, in relation to $\sqrt{\lambda C}$ for the finger. Our concern in the present context is the property in relation to soils and rocks. But it must be remembered that the air also has a conductive capacity coefficient – which is somewhat variable.

The emphasis on the exchange of heat in the immediate vicinity of the surface, must not lead to the conclusion that the value of the conductive capacity coefficient at the surface only is important. The exchange at the surface is influenced by the conductive capacity at all depths affected by the surface temperature disturbance. Thus variations in soil material with depth become relevant.

Discussion of this derived thermal property of soils which relates directly to energy exchange at the surface serves as an introduction to the more detailed considerations of ground climate, and of surface heat exchange, in Chapters 8 and 9.

6

Soil-water relations

6.1 The significance of water in earth materials

Soils normally contain a variable amount of water, often more by volume than of solid components. This liquid component has a fundamental significance for the properties and behaviour of the material. The importance of water in relation to certain mechanical properties was considered in Chapter 4.

The soil water is largely responsible for the availability of plant nutrients and the quantity and condition of the soil water is a major ecological factor. Natural processes such as weathering and soil movement on slopes, and properties such as the degree of compaction and the consistency or strength of the material cannot be considered without full regard to the amount and state of the water. Near the surface of the ground variations of water content occur constantly in association with weather on a day-by-day, or seasonal pattern. A total, or a periodic long-term lack of water so affect both properties and processes as to give rise to distinct landforms (those of the arid areas, or of the surface of the moon, are extreme examples). Generally speaking, the presence of water greatly complicates analyses of the behaviour of earth materials, compared to the same materials in a totally dry state.

Specialists in different disciplines tend to analyse the role of water in different ways. On a practical level.the farmer is essentially concerned with the availability of water in the near-surface layers where complex forces of retention are dominant. The hydrologist and foundations engineer are commonly more concerned with materials below the water table where water pressures are greater than atmospheric so that relatively simple considerations of hydraulics and hydrostatics may suffice. Modern technology as well as the requirements of a truly scientific approach, however, necessitate a more comprehensive understanding.

Water in soils occurs primarily in small spaces – the pores – and is therefore often exposed to a very large area of mineral or other solid surface. This situation and as a result the state of the water, is special, as mentioned in Chapter 2, and an elementary understanding of hydrostatics and hydraulics of flow within pipes or other large volumes is insufficient in this respect. Many of the considerations of fluid mechanics as applied in studies of rivers or lakes, are not directly relevant to the basic understanding of water in porous media.

A misconception along different lines, which has persisted from the earliest studies of soil water, is that the *amount* of water present is directly responsible for whatever properties the water may impart to the soil. It is on the contrary, the *energy* status of the soil water that is more often significant. The use of the word 'energy' in this context is discussed in detail later in this chapter. Of course the energy status of the water in a given material changes as the water content changes, but the energy status associated with a particular water content depends on the nature of the soil particles, their size and arrangement, and also on the situation of the soil being considered.

In the following sections, various physical relationships are described which explain these dependencies. Because the phenomena are to a large extent microscopic, or even sub-microscopic, the physical descriptions are usually incomplete. However, in directing attention to energy relationships we are also able to employ the methods of thermodynamics which permit a deeper understanding of the relationships. The thermodynamic approach reveals general relationships of wide applicability, this is also so in cases where phase change occurs and where our strictly physical knowledge of the soil's nature and behaviour may be limited.

To illustrate the foregoing, consider the cases of two soils, a clay and a sand, each with 20 per cent water content. The sand may well be saturated. On squeezing, a small quantity of free water will then appear. The sand is soft and easily separated, it has low strength. Plants would grow in it (assuming nutrients are present) or at least remain turgid. The clay on

the other hand may well appear quite dry. Depending on the mineralogical composition it will be a stiffly plastic, or semi-rigid material, and thus show considerable strength. Plants will in all probability wilt rapidly if rooted in such material. If the two materials are left in intimate contact for a period of time, water will be found to have left the sand and entered the clay. Finally, if left exposed in a room, careful measurements may show the sand to be losing water to the atmosphere by evaporation, while the clay is gaining water by condensation.

The significance of the nature of the soil for the behaviour of soil water cannot be over emphasized. Even relatively slight textural or compositional differences affect the energy status corresponding to a particular water content. Thus comparisons of water content as a guide to soil properties are only fully meaningful if the soils referred to are one and the same material. If nothing is known of the nature of the soils except their water contents no conclusions can be drawn.

6.2 Capillarity

If a glass capillary tube is placed vertically with its lower end in water, water is seen to rise up the tube to a level above that outside the tube, Fig. 6.1. It will rise higher in a tube of smaller internal diameter. The height of the rise h is given by

$$\rho g h = \frac{2 S \cos \theta}{r} \qquad [6.1]$$

$$\text{or } h = \frac{2 S \cos \theta}{r \rho g}$$

where S = surface tension air-water
r = radius of capillary
ρ = density of water
g = gravitational acceleration
θ = contact angle. In Fig. 6.1 θ is $180\,° - \alpha$, for clean glass and pure water $\cos \theta = 1$

This so-called capillary rise is due to surface tension which is a property exhibited by interfaces between solids, liquids or gases (Sears and Zemansky 1965). Surface tension is

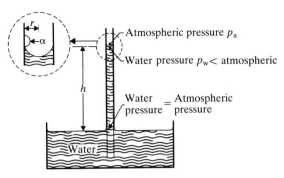

Fig. 6.1 Rise of water in capillary tube

expressed in dimensions of force per unit length, and its value depends on the substances involved across the interface. For water-air interfaces it is 72.75 mN m^{-1} at 20 °C (Kaye and Laby 1973). It has its origin in molecular forces which are not balanced when the interface is small, until equilibrium is obtained by the occurrence of different pressures across the interface. In the case of the capillary tube described the pressure of the water below the interface (or 'meniscus') is lower than atmospheric, and the difference is referred to as its suction. Attainment of equilibrium requires that water move upward from the container where it is at atmospheric pressure, to give a column of water of height h. The effect of the weight of the column (expressed as a pressure $\rho g h$ – compare p. 48), is such to exactly balance the suction effect, $\dfrac{2 S \cos \theta}{r}$

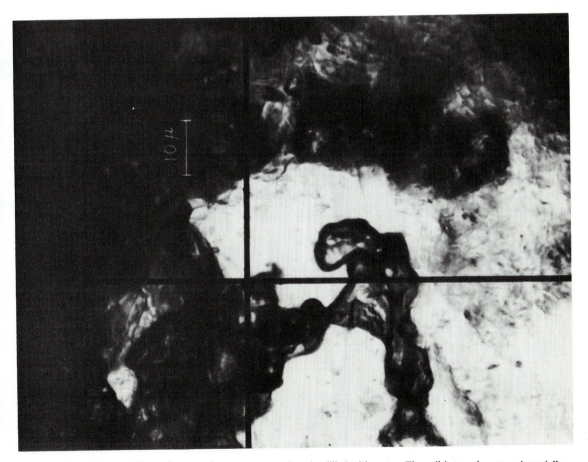

Plate 6.1 Air (dark areas) in a soil, occupying pore space otherwise filled with water. The soil is translucent and specially prepared from the mineral cryolite. The lighter shadows are impurities in the mineral particles which are otherwise largely invisible. The curved (and in some cases clearly semi-circular) boundaries are the air-water interfaces (menisci) – associated with suction in the soil water (from Williams, 1967). $10 \mu = 10^{-6}$m

Soil pores usually interconnect and although they do not constitute uniform cylindrical spaces, a suction is developed where there are interfaces between water and air. It is sometimes referred to as soil water tension. Capillary menisci in soils are shown in Pl. 6–1. The suction causes the capillary rise which is responsible for the occurrence of substantial amounts of water in the soil at levels above the water table. The latter is defined as the level at which water in the ground is at atmospheric pressure, and at which water would stand in an open pipe in the ground (Fig. 4.3).

As the pores of a soil have a range of sizes there is no single height of capillary rise for a soil. Rather, depending on their size, the number of pores filled with water decreases with height above the water table. Sand has many pores of radii 0.001 cm–0.01 cm (often called equivalent radii, because the pores are irregular in shape). Consequently the zone which is more or less saturated with capillary held water extends only some centimetres or decimetres above a water table (in accordance with [6.1]). Clays have pore radii in the micron (10^{-4} cm) range and smaller, which implies a capillary rise of tens or hundreds of metres. Actually the smallness of the pore openings constricts the rate of water movement, and water will not *rise up* a 100 metre high column of clay in any practical period of time. Nevertheless, water in fine-grained soils experiences suction, often to a very great value.

Consider a laboratory experiment in which a column of soil, say fine sand, is placed with its lower end in water as in Fig. 6.2, and left for sufficient time that the water has ceased to move. If samples are then removed from the column, it is found that their water content is smaller (fewer pores being filled with water) as the height above the surface of the free water was greater. The pressure of the water in each sample is also lower, as the height h of the sample was greater. The height above the free water is a measure of the suction (which can be thought of as the negative or sub-atmospheric pressure in this case), with the units 'cm water column'. A graph may be drawn and this is an example of a suction-water content relationship or soil moisture characteristic, of which a number are shown in Fig. 6.3. The units 'm water column' may be converted to a pressure, by multiplying by ρ_w, the density of water, kg m^{-3}, and g the acceleration due to gravity 9.81 m s^{-2}.

If samples are taken from two different heights in the column and placed in contact, water will move in to that sample with greater suction. Suction is a measure of potential of the soil water, and it is differences of potential which cause water movement. However, suction is not, as we shall see, the only source of potential, such that the water *in the vertical column* is in equilibrium in spite of the suction gradient. The suction is counterbalanced by another potential.

6.3 Determination of suction-water content relationships

Practical limitations on the height of the soil column generally prevent the procedure outlined in Section 6.2 being useful in practice, for studying these relationships. *Suction plates* (Croney, Coleman and Bridge 1952) may be used, Fig. 6.4a. These are porous ceramic plates to the underside of which is attached a hanging tube of water. When a wet soil sample is placed on the plate, water leaves the soil passing through the plate into the tube until a suction develops equal to that represented by the hanging water column. At that time capillary menisci will have developed in those

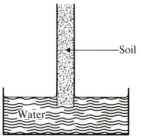

Fig. 6.2 Soil column with free water at base

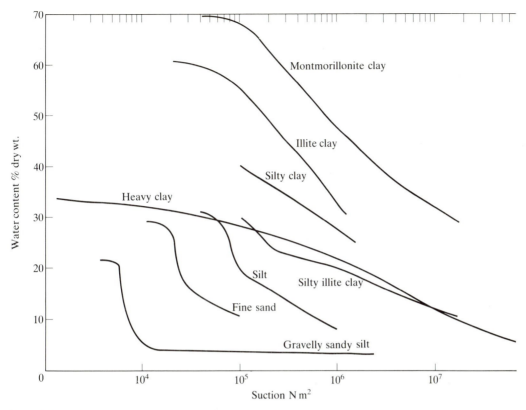

Fig. 6.3 Examples of suction water content relationships, drying curves (see text)

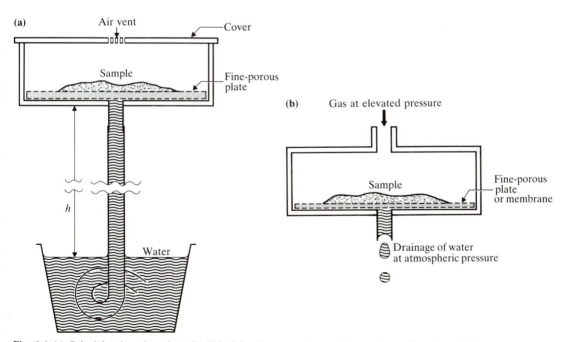

Fig. 6.4 (a) Principle of suction plate **(b)** Principle of pressure plate and pressure membrane apparatus

soil pores of a size, according to [6.1], corresponding to the length of the water column.

The length of the water column can be varied somewhat but obviously cannot be very long. The range of suctions obtainable can be greatly increased by a modification, the *pressure plate apparatus*. To understand this, we first note that an alternative, and more general equation to describe the surface tension effect ([6.1]) is:

$$P_a - P_w = \frac{2 S \cos \theta}{r} \qquad [6.2]$$

Here P_a is the pressure of the air above the meniscus and P_w, the pressure of the water just below the meniscus. Written in this form, the equation emphasizes that the surface tension is responsible for a *difference* in pressure between the phases, and this difference, $P_a - P_w$ is the suction. At depth in the ground, air in the soil may be at above atmospheric pressure, and on occasion the soil water may also be, while a state of suction exists. In the pressure plate apparatus, Fig. 6.4b, the sample is enclosed in a container the bottom of which is the plate, and into which air (or nitrogen) may be passed under pressure. The tube is greatly shortened and is simply a drainage outlet. The situation is now as described by [6.2], with an elevated gas pressure and water from the soil draining to that in the short tube at atmospheric pressure.

With either device, when moisture equilibrium is reached, the moisture content of the sample is determined and plotted as in Fig. 6.3. Common to all such devices is the role of the porous plate. The pores in the plate must be small enough that they are water filled at the applied suction ([6.2]). If there are continuous pores through the plate, of an equivalent radius greater than that given by [6.2] for the applied pressure difference, these pores will immediately drain into the tube below and air will follow rapidly outward through the plate.

The functioning of the porous plate illustrates the behaviour in the soil itself; when the suction is further increased additional

smaller pores empty. As a soil dries the water remaining is confined to smaller and smaller pores.

The suction-moisture content relationship can be extended to much lower moisture contents by the use of cells able to withstand air pressures up to hundreds of times greater than that of the atmosphere. Clearly a very small-pored material is then necessary for the base 'plate' and various kinds of cellulose membranes are used. In this case the apparatus is known as a 'pressure membrane' apparatus.

6.4 Attributes of the suction-water content relationship

Soils which are not saturated have some air in their pores, and thus have confined air-water interfaces. Therefore, according to [6.2] the water is in a state of suction. Suction may also occur in saturated soils. Consider a sample of clay, absolutely saturated and 'wet'. If water is then lost this will be from the surface, resulting in the formation of menisci there. Because all the pores are very small the menisci will (initially) have radii bigger than pore size and will therefore be restricted to the surface. Only when a sufficient suction has developed, will the menisci radius be as small as that of the largest continuous pores, which can then empty. The suction value at which this happens is the air entry or air-intrusion value (Williams 1967). Development of suctions less than the air intrusion value is associated with consolidation of compressible soils.

So long as the soil is saturated the suction constitutes an equal effective stress $\acute{\sigma}$ (see p. 48). As shown by [6.2], suction is a difference in pressure between air and water. The air pressure on a sample constitutes the total stress, σ, while the water pressure is the pore pressure U (compare section 4.7). Thus

$$P_a - P_w \equiv \sigma - U = \acute{\sigma} \qquad [6.2a]$$

(In the ground there is usually an additional component of total stress from the weight of the overlying soil.)

Because the word 'suction' colloquially has connotations of sub-atmospheric pressures, of the drawing of water up drinking straws, or from wells by suction pump, two special attributes of soil water suction must be considered. Firstly, as discussed in relation to equation 6.2, when the air pressure is above atmospheric, a state of suction may exist in which the water pressure is also above atmospheric. Secondly, there is no constraint on the magnitude of soil water suction similar to that preventing water being 'sucked' up a pipe to more than about 10 m height. Elementary physics teaches that water has no tensile strength and it is in reality the air pressure which pushes the water up to a maximum of this amount. But in the case of finely-porous media, suctions equivalent to columns of water of hundreds of metres occur. Such suctions often involve to a considerable extent 'negative pressures' in the absolute sense and can thus be regarded as a state of tensile stress in the water. Although water in bulk cannot be pulled and a column breaks because the water has no tensile strength, when water is confined in very small spaces this is no longer the case. Earlier the subject of scientific controversy it is now well established that the conditions for initiation of rupture, the nucleation of small, usually sub-microscopic cavities or bubbles frequently do not exist in small, confined water masses. A pore does not empty unless a certain suction exists, and a vapour bubble (or pore) of smaller size is immediately refilled by water (Williams 1967).

The existence of tensile stress effects and the adhesion of the water to particle surfaces can give great strength to soils (See chapter 4). The suction serves to pull the particles together. The strength of a sample of moist silt, compared to its loose state when dry is well known. Both the atmospheric pressure and an eventual tensile stress in the water contribute to the effective stress $\acute{\sigma}$, which directly affects soil strength (see p. 51).

The suction is often referred to as the *soil* suction. Although the suction is in the water, the magnitude of the suction is so closely related to the nature of the soil that this usage is unobjectionable. Workers in some fields refer to the suction water content relationship as the soil-water retention characteristic. This is also meaningful in that the suction is a measure of the energy required to remove water from the soil. A plant must generate a greater 'suction' (more precisely, a lower potential) than that existing in the soil if water is to move along a gradient from the soil into the roots.

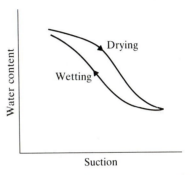

Fig. 6.5 Hysteresis in suction – water content relationship

Suction-water content relationships (Fig. 6.5) exhibit hysteresis (or 'partial irreversibility'). The term here refers to the fact that the moisture content for a specified suction is somewhat greater if the moisture content is arrived at by a process of drying as opposed to wetting. There are several reasons for the phenomenon. For example the contact angle Θ (assumed usually to be $0°$ Fig. 6.1) may be greater on wetting (compare how spilt water spreads with a frontal 'ridge', which disappears when the water is drawn back by an absorbent tissue) and the suction is changed. Similarly on 'drying' water may remain in a pore longer than implied by [6.2] if the entrances to the pore are all smaller than the pore itself and remain water-filled, effectively blocking entrance of air.

Strictly speaking, therefore, there are a range of possible moisture contents corresponding to a given suction, depending on the extent of wetting or drying that has occurred. In practice, most suction-water content relationships are determined on initially saturated specimens and are thus 'drying' curves. The application to practical problems of values so obtained may occasionally involve significant inaccuracies, and the possible significance of hysteresis must be borne in mind.

At low suction, as occurs for example in a sand which is wet but not saturated, the measured suction as far as most of the soil water is concerned is simply a state of pressure lower than that of the surrounding air, due to the capillary effects discussed. At high suctions the water in the soil is restricted to smaller spaces, and, is in fact, largely in the form of thin layers adsorbed on the particle surfaces.

The state of water in such adsorbed layers has been discussed briefly in Chapter 2. Although the matter is very complex it seems clear that because the water molecules are under the influence of forces occurring normal to the particle surface, it is at best incomplete to view such water as merely experiencing a lowered pressure. Indeed one would rather suppose that the water molecules were being pulled together on to the particle, and thus in a state of elevated pressure. It must be accepted that our usual understanding of 'pressure' (which derives from consideration of water in bulk, or 'free' water) is somewhat irrelevant. As the water content is reduced, the measured suction increases in a regular fashion to as low a moisture content as we care to consider. This further suggests that suction is a more fundamental property and that the term is more comprehensive, than its conventional literal meaning. Schofield (1935) proposed that suction as in a suction-water content relationship, was better regarded as a measure of the Gibbs free energy (see p. 24). Although the proposal requires qualification, it introduced a most valuable concept: that there is a unifying quantity which can be used to measure the tendency of movement or transfer of water from one position to another in the ground even though the situations of the water at the two positions might be different in a number of respects.

The expressions 'free water' and 'bulk water' appear quite often in our discussion. As

discussed in Chapter 2, p. 18, they are used with reference to properties of water when the water is free of the effects of microscopic surface forces. Water in soils is frequently neither 'free' nor 'bulk', the latter being also true of, for example, water in tiny droplets in the atmosphere. The thermodynamic approach to soil moisture is largely a quantifying of the differences between 'ordinary' free, bulk water under atmospheric pressure, and the soil water. Free energy is an example of a thermodynamic function convenient for the purpose.

6.5 Energy and potential

It was pointed out in Chapter 2, that a condition of equilibrium in a system is that the Gibbs free energies, G, of the components are equal. When inequality of G exists between the components change will tend to occur in such a way as to produce equilibrium, with a minimum value of G. The use of the free energy function is by no means limited to the study of the direction or rate of chemical reaction. The values of free energy of two phases, for example, ice and water, are equal only at the freezing point (more strictly speaking, we should specify it is the free energy *per mole*, or *per gram*, because of course a little ice might be in equilibrium with a lot of water). At lower temperatures, ice, which has the lower free energy, is formed spontaneously from the water. Because free energy is a function of pressure and temperature, as well as the substance (water or ice) the freezing point can vary.

In the case of soil water, it is differences in free energy which are responsible for the flow of water from one point to another. The speed of the flow depends on the gradient of free energy. In a more elementary view of liquid flow, of course, we immediately think of pressure differences as causing flow, and, indeed, if the pressure difference is the *only* cause, it can be equated with the free energy difference. But the importance of the free energy function lies in its comprehensive nature, enabling us to establish associations between chemical, mechanical and thermodynamic properties and processes. The identification of suction with Gibbs free energy opens up a broad vista of relationships highly significant in our understanding of soil water. These relationships were the subject of a classic monograph by Edlefsen and Anderson (1943).

In view of all this, it may be surprising that the term soil water potential is more usually met with in the study of soil water movement (Day, Bolt and Anderson, 1967). In many respects the term is equivalent and the two quantities free energy and potential will commonly be equal numerically. However the Gibbs free energy is ultimately defined in terms involving heat and temperature (p. 124), while 'potential' as used in soils studies is strictly speaking a quantity applicable in problems of a mechanical or hydrodynamic nature, where temperature change is not involved. The Gibbs free energy will be utilized in discussion of soil freezing (Chapter 7), and the correspondence of the two quantities will be considered further in later sections.

The concepts of free energy and potential provide a logical framework to describe the behaviour of water in soils. If we attempt to describe this behaviour solely in terms of analogies with the behaviour of water in macroscopic tubes or pipes where pressure gradients exist we limit our understanding and indeed may on occasion arrive at the wrong conclusion. Instead, the problems can be considered better in terms of energy or potential differences.

The particular symbols and definitions used in the following discussion, are as far as possible in accordance with those proposed by the International Society of Soil Science (ISSS, 1963, 1975). The existence of a potential gradient implies a tendency to movement or flow according to an equation of the type:

$$q = k\nabla\psi \qquad [6.3]$$

where q represents the flow or flux or water, k is the coefficient known as the hydraulic conductivity or permeability which represents the ease with which the water may travel through the particular soil and $\nabla\psi$ is the gradient of potential or change of potential per unit length.

The similarity of [6.3] with, for example, those used in studies of fluxes of thermal or electrical energy will be recognized. In this respect, temperature is analogous to potential. Although temperature can be measured on an absolute scale (K), we frequently use a scale with a chosen datum (e.g. 0 °C on the centigrade scale). It is not possible to measure free energy or potential on an absolute scale, and instead we define for each case, a datum against which we measure the quantity. Normally, therefore one understands that relative potentials (although often just called 'potentials') are involved, and in every case it is essential to understand the nature of, and define clearly, the datum.

In the suction plate procedure, the water in the soil is brought to equilibrium with (that is, has the same potential as) water outside the plate at some 'sub-atmospheric' pressure. Thus it is understood that the datum was in fact atmospheric pressure – or more precisely a body of free water at that pressure and at the same elevation as the soil sample. The suction so measured (in units of pressure or in m of water column) corresponds to the soil water potential and is known as the matric or capillary potential (with units J kg^{-1} – we shall see later how the units correspond). The latter is only one of several potentials which relate to water in soils and the tendency to movement depends on a total soil water potential which is often, indeed almost always, the sum of a number of component potentials. The definitions and units, and interrelationships of suction, potential and free energy are reviewed in Aitchison, Russam and Richards (1967).

6.6 Matric potential

Following ISSS (1963) the formal definition of matric potential is: 'the amount of work that must be done per unit quantity of pure water in order to transport reversibly and isothermally an infinitesimal quantity of water from a pool containing a solution identical in composition to the soil water at the elevation and the external gas pressure of the point under consideration to the soil water'. In this apparently unwieldy definition we first note that the opening words are simply a definition of an energy difference between the soil water and the datum water body. Energy is the capability of doing work (section 1.2) and a particular energy difference can be equally well measured in terms of the amount of work involved in establishing it. The reader should not concern himself with whether the work done might be 'positive' or 'negative'. We are concerned only with increments of potential, the sign of which will be determined by our choice of datum and the circumstances of the soil water in question, as will be discussed later. Capillarity and adsorption are the source of forces associated with matric potential. The remainder of the definition is concerned with specifying the relationship of the soil water and the datum water body, in measuring this potential. The datum water body is 'a *solution* identical in composition to the soil water' to exclude any effect due to osmotic differences. In practice this requirement is often ignored without significant loss of accuracy.

6.7 Gravitational potential

The reference to elevation in the definition of matric potential emphasizes that a difference in elevation itself constitutes a potential – water tends to move from higher to lower elevations. Such a component potential is called the gravitational potential. It is a result of the forces due to gravity and is given by $\rho\,g\,h$, where h is the height difference (that is the height above the datum), g is the gravitational acceleration and ρ the density of water. In the experiment with the vertical soil column (Fig. 6.2) equilibrium was attained when the

demonstrated 'suction' or matric potential gradient was exactly balanced by the gravitational potential gradient.

If the column of soil is turned on its side the gravitational potential gradient is removed. There will immediately commence a flow of water along the suction gradient, which continues until the moisture content becomes uniform. Conversely, although soil at two points on a slope may have equal matric potentials, flow will still tend to occur because of the elevation difference.

It should be noted that hydrologists and engineers frequently use the term head in dealing with this kind of question, and in a sense essentially similar to that of potential. Provided the same attention is given to definition the term itself is unobjectionable. Clearly the gravitational potential is important in analyses of ground water flow in aquifers and adjacent regions (see e.g. Ward 1975).

6.8 Osmotic potential

A further component potential, the osmotic potential, exists because of dissolved salts in the soil. In identifying the suction with matric potential for the experiment with the soil column, salts were assumed absent. In fact they commonly occur in solution to some degree. Solutions have a tendency to draw pure water to themselves when separated from such water by a semi-permeable membrane. Such a membrane is one preventing the passage of the solute molecules, while allowing water to pass. 'Osmotic pressure' refers to the lower pressure which appears to exist in the solution than in the pure water (Fig. 6.6). The osmotic potential is not properly accounted for in the suction plate test. While the pores of the plate are small enough to prevent passage of air (because of the capillary effect) they do not present a barrier to salts. Thus during the suction test dissolved substances (which may have broken down to elemental ions) pass easily from the soil into the water, in and below the plate.

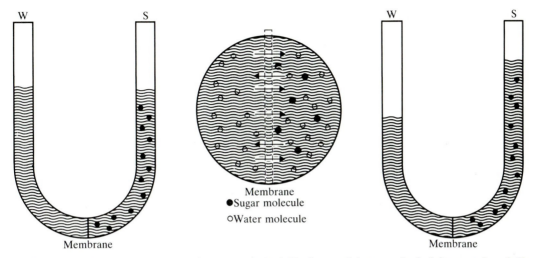

Fig. 6.6 Illustration of osmosis and of osmotic pressure. Left: A U-tube containing water in the left arm and a solution of sugar in the right arm. These are separated by a membrane permeable to water molecules but not to dissolved sugar. Centre: Enlarged portion of the membrane with H_2O molecules moving freely from the water side to the solution side and vice versa, while the sugar molecules are unable to penetrate the membrane. Right: the effect of sugar is to decrease the free energy of the water on the solution side and more water therefore passes from left to right than from right to left. When equilibrium is reached the difference in water level in the two arms represent the 'osmotic pressure'. (After Brady 1974)

Were there to be a truly semi-permeable membrane there would instead be a holding back of water in the soil due to the attractive forces of the dissolved molecules.

It is difficult to assess the role of the osmotic potential in soil water movement. Although semi-permeable membranes in the sense described obviously do not occur in soils naturally, the attraction the surface of particles exerts upon cations results in these to some degree being held on to the particle surfaces. The ions are not then free to move out of the soil, and instead will themselves tend to hold, or draw water into, the soil. In most soils other than certain clays it is likely that the osmotic potential is of little significance in causing water movement (Rose 1963). It is important for the movement of the ions or dissolved compounds themselves into other parts of the soil by diffusion. In certain clays, however, the effect of osmotic forces originating in the surface region of the particles is very marked. These, so-called swelling clays, belong to the montmorillonite clay mineral group, and are characterized by an almost insatiable attraction for water which continues to move into them

causing steady increase of volume (see also p. 54). Application of substantial loads to the clay may be necessary to counteract this tendency. The load required itself constitutes a measure of the osmotic potential.

6.9 'Tensiometer potential' and additional component potentials

A device known as a tensiometer is commonly used for measuring potential of soil water in the ground. It operates on the same principle as the suction plate, involving a water column in contact with a fine-porous cylinder (Fig. 6.7), the outer surface of which is in intimate contact with the soil. The cylinder, and a tube leading to the ground surface are filled with water. At the top there is a U-tube containing mercury, or a Bourdon gauge capable of measuring sub-atmospheric pressure, or other pressure transducer. The apparatus has the same limitation as the suction plate, that is the measured pressures or suctions can only be positive absolute pressures. In practice they cannot be less than about 0.4 times the pressure of the atmosphere (i.e $< -0.6 \ 10^5$ N m^{-2}, if the pressure of the atmosphere is taken as zero), because the water in the water column then ruptures.

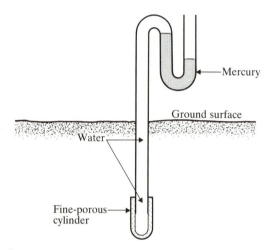

Fig. 6.7 Principle of tensiometer

Tensiometers are used in field measurement to determine irrigation requirements. Because the availability of water to plants depends on potential, the tensiometer results are more directly applicable than measurements of water content of the soil. Practical limitations of tensiometers means that irrigation requirements are also estimated from water content measurements using various devices (Withers and Vipond 1980). But such measurements must then be interpreted, with assessment of the suction water content relationship for each soil. It is important to understand the so-called 'tensiometer potential' which a tensiometer measures. It does not measure exactly the suction that would be given by the suction-water content relationship for a sample taken from the soil. It usually approximates it, but the distinction is fundamental. The tensiometer shows the pressure of the water in the porous cylinder, which is in equilibrium with that in the soil. The gauge of a tensiometer measures this water pressure

relative to the air around the gauge, that is, relative to atmospheric pressure (a correction is made for the length of water column between the gauge and the cylinder).

Now the suction, as measured in the suction-water content test is $P_a - P_w$ (cf. [6.2]), it being understood that P_a refers to the pressure of the air adjacent to the air-water menisci in the pores of the soil. The air pressure P_a in the ground may, under special circumstances, be higher or lower than atmospheric, but this cannot be assessed from the tensiometer reading alone.

The matter may be clarified by considering two samples of the same soil, one of which is placed on a suction plate and a certain suction applied, while the other is placed in a pressure-plate apparatus where the same suction is applied by raising the air pressure. Although the samples will assume the same water content, and have the same suction it is also clear that if tensiometers could be inserted in the soil samples they would give different readings, in fact they would show the pressures of the soil water in each case.

It is also clear that if suitable connection could be made between the two samples, water would flow from that sample on the pressure plate to that on the suction plate. The additional component potential associated with an elevated air pressure in the soil pores is sometimes called the *pneumatic potential*. It is a component of the tensiometer potential.

An additional point relative to 'tensiometer potential' is that the application of overburden (additional soil or other weight on the ground) is often associated with a rise in the measured pore-water pressure (or potential). For coarse and incompressible soils, the effect is usually small, there being only a short-lived small rise in the water pressure which is rapidly dissipated by flow. In saturated compressible soils, however, there occurs the highly significant consolidation process, which was discussed in Chapter 4. There is an immediate rise in observed water pressures, by an amount equal to the applied overburden pressure. This increment of pressure is dissipated over a long period. As an element of potential it is referred to as the *envelope pressure potential*, but it has been proposed (ISSS 1975) that the envelope pressure potential be regarded merely as a sub-component of the matric potential. However it be classified, it is a quantity of great importance in analysis of the consolidation process. Geotechnical engineers refer to it as the hydrostatic excess pressure, a term which emphasizes transience. Once hydrostatic conditions are restored, there is no envelope pressure potential.

An envelope pressure potential is included in a measured tensiometer potential. Indeed, in geotechnical engineering, the process of consolidation is often monitored by piezometers – simple devices which differ in only one essential respect from tensiometers. They do not include a fine-porous filter to prevent entrapment of air, as they are intended for use in saturated soils. The positive pressure measured is referred to as the porewater pressure (symbol U) and in other respects corresponds to the tensiometer potential. Similarly a value of $U = O$ corresponds to atmospheric pressure, as does a tensiometer potential equal to O. Changes of U associated with loading of the ground represent envelope pressure potential.

6.10 Total potential in the ground

When considering movement of water in the ground one must consider differences in total potential. Total potential may be written as:

$$\psi_t = \psi_m + \psi_g + \psi_o + \psi_p + \psi_{env} \qquad [6.4]$$

In a particular case not all the terms may prove relevant. There may be no envelope potential ψ_{env}, or no pneumatic potential ψ_p. The osmotic potential ψ_o may be insignificant, or ineffective with respect to water movement. If the points of interest are at the same elevation there is no gravitational potential ψ_g.

If the soil is saturated it might be thought that because there were no capillary menisci

there would be no matric potential. This is not correct because the important effect of hydrostatic pressure due to depth below the water table is also included under matric potential. There are good reasons for doing this even though the term 'matric' may appear to be rather over-extended. The hydrostatic effect is $\rho\,g\,h$ – and this applies equally well at a depth h below the water table in the ground, as at a depth h in a swimming pool. It is of course the weight of water that produces it. In the absence of envelope pressure potential and pneumatic potential, $\rho\,g\,h$ is under static conditions the pore-water pressure, U. In a soil column under these conditions U increases *downwards* in a linear fashion, from less than atmospheric pressure above the water table where it corresponds (cf. section 6.2) to the suction, to increasing positive pressures below the water table. Clearly the porewater pressure can then be identified with the matric potential above the water table and logically therefore also below it.

Care should be taken not to confuse the hydrostatic pressure effect with the gravitational potential which exists because of elevation differences: clearly the latter increases in an *upwards* direction. The two potentials are numerically equal but of opposite sign. They therefore balance out and this is why, although there is a pressure gradient with depth in a swimming bath, there is no movement associated with it. In Fig. 6.8 potentials in soil columns under various circumstances are shown.

6.11 Units and dimensions

It has been stressed that the tendency for transfer of water (whether from one position in the ground to another, from one phase to another, or from one state of chemical combination to another) depends on energy differences. As a measure of energy difference, potential ψ has the units joules per kilogram (as does the Gibbs free energy). But

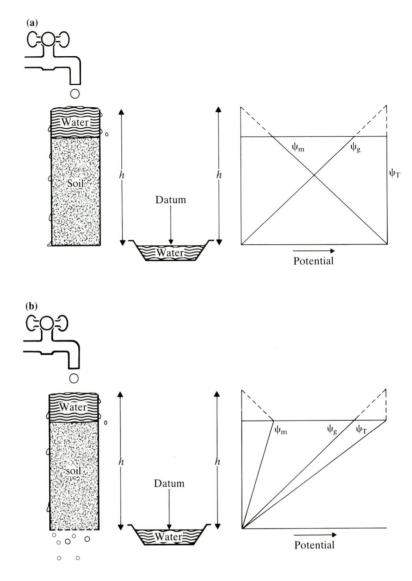

Fig. 6.8 (a) Soil water potentials. The soil is a saturated salt-free sand. The matric potential m increases with depth (ie. decreases with height above datum). The gravitational potential ψ_g increases with height above datum. The sum $\psi_m + \psi_g = \psi_T$, the total potential, is constant, and the water in the soil is therefore at equilibrium. (After Bolt 1970)
(b) Here the gradient of matric potential ψ_m (in effect a pressure gradient) is dependent on the permeability, which controls the drainage through the column following from the gradient of total potential ψ_T. (After Bolt 1970)

in considering flow of water through a soil we are normally concerned with volumes of water which will pass along a certain path of a certain cross-sectional area, for example cubic metres passing through a square metre in a second. Consequently the potential, as will become clearer in the following section, may be required per unit of volume, rather than of mass. If the potential expressed as J kg^{-1} is ψ, then $\dfrac{\psi}{\bar{V}}$ gives us J m^{-3}, where $\bar{V} = \dfrac{1}{p}$, the specific volume, or cubic metres of water per kilogram. $\dfrac{\psi}{\bar{V}}$ is also called potential (it might avoid confusion were it known as potential per unit volume), and when the cgs system of units was used $\bar{v}$ was simply 1 cm^3 gm^{-1}. In SI units $\bar{v}$ is 1 10^{-3} m^3 kg^{-1} and care must be taken not to overlook the factor of 10^{-3}.

A Joule is equivalent to a Newton metre so that:

$$J\ m^{-3} \equiv N\ m\ m^{-3} \equiv N\ m^{-2}$$

The newton is the unit of force, and N m^{-2} is therefore force per unit area – which are the dimensions of pressure. We see that $\frac{\psi}{}$ can be regarded as a pressure, and indeed hydrologists and engineers commonly refer to pressure (as in 'porewater pressure') in considering water movement in soils and rocks.

The reader may wonder why it is necessary to have such lengthy discussion of potentials as energy differences rather than just referring to 'pressure'. There are several reasons. What we really mean by porewater 'pressure' is that the state of the soil water is such that it would be in equilibrium with (have the same potential as) pure free water subjected to that pressure. The potential ψ per unit of mass (e.g. J kg^{-1}) of pure free water due to pressure alone is s mply $\Delta P\ \bar{v}$. Even more simply, on a unit volume basis it is equivalent to ΔP, the pressure, in e.g. N m^{-2}. The specific volume of water $\bar{v}$ can be taken as constant. But the water in the soil is neither pure nor free (by free we would mean that its properties are not affected by the intrinsic nature of the container), and a number of the component potentials concern effects that cannot be regarded in the strict sense as 'pressure'. If we had a tiny pressure gauge normal in every respect except that it could be placed right inside the pore, quite frequently it would not register the pressure actually indicated by an adjacent tensiometer. Indeed, the pressure might vary with location in the pore. But such variations are compensated by other component potentials (osmotic or adsorption effects) such that there is presumably a uniform total potential. The balancing potential in the tensiometer (after equilibrium is reached) is, however, due solely to pressure in the normal sense. Table 6.1 shows equivalent measures.

It is only by consideration of all the possible component potentials that we can be sure of considering all the phenomena which determine the tendency to movement, or understand the distribution of moisture content. In Chapter 7 where the effects of freezing are considered it will also be apparent that difficulties can arise if we take the simplistic view that the depressed freezing point of soil water is solely due to pressure effects.

The fact that the tensiometer and piezometer give measurements in the form of equivalent pressures, facilitates the use of simple equations to determine the magnitude and direction of the soil water movement. It remains important to remember exactly what the tensiometer and piezometer are doing, and two examples will suffice to show how uncritical use of such devices can lead to error. If a soil is strongly saline, pure water moving out of a tensiometer may reduce the osmotic potential significantly by dilution, while producing expansive pressures in the soil with a chain of side effects influencing the 'equilibrium' tensiometer pressure. As a second example in cold climates tensiometers are sometimes filled with ethylene glycol solution as anti-freeze. Not only is the water in the tensiometer not pure, but the ethylene glycol diffuses into the soil where for example, it reduces surface tension and thus the matric potential. The problem can be overcome by a technique involving a short length of mercury at the bottom of the tensiometer tube, thus separating the glycol solution from water in the tip and soil. The tip itself must be in unfrozen soil.

6.12 Flow equation; practical procedures

An equation of the type of [6.3] is:

$$q = k\ i\ A \qquad [6.5]$$

where the rate of flow or transport is q. The cross section through which flow is occurring is

Table 6.1 Some equivalents* used in the measurement of suction, potential† and pore pressure†

cm water column	$J\ kg^{-1}$	$Kgf\ cm^{-2}$	Bar	$Nm^{-2}\ (=Pa\ \ddagger)$
1	$9.81\ 10^{-2}$	$1\ 10^{-3}$	$9.81\ 10^{-4}$	$9.81\ 10^{1}$
10	$9.81\ 10^{-1}$	$1\ 10^{-2}$	$9.81\ 10^{-3}$	$9.81\ 10^{2}$
100	9.81	$1\ 10^{-1}$	$9.81\ 10^{-2}$	$9.81\ 10^{3}$
1 000	$9.81\ 10^{1}$	1	$9.81\ 10^{-1}$	$9.81\ 10^{4}$
10 000	$9.81\ 10^{2}$	$1\ 10^{1}$	9.81	$9.81\ 10^{5}$

* The values are based on a density ρ of water of 1 g cm^{-3}.
† The values coincide provided the datum for potential involves atmospheric pressure, and if the pore pressure is also expressed relative to atmospheric pressure.
‡ The pascal (Pa) is an accepted SI unit for pressure, although not used in this book (1Pa = 1 N m^{-2}).

represented by A. It is essential that rational and compatible dimensions and units of measurement are used for the various terms. We have seen that potential can be expressed in more than one way (per unit mass, per unit volume or as a pressure) and consequently so can the gradient of potential here represented by i. The choice of units for this term determines the appropriate unit for the transport coefficient k, which in the case of water movement is usually known as the permeability or hydraulic conductivity. Together k and i determine the units of q.

In practice it is frequently convenient as noted to consider the potential as a pressure. In considerations of flow this is often expressed as height of a water column. We visualize a column of water of unit cross-section and height h, the weight of which produces a pressure given by ρgh. As we can usually consider ρg as constant, h itself serves to measure the pressure. It does not matter if part of the column narrows, even to a cross-section of less than unit area. A vertical column of water of height 10 m exerts a pressure on a horizontal square metre at its base of: $\rho gh =$ 9.81 10^4 with the units kg m^{-3}m s^{-2} m that is, kg m s^{-2} m^{-2}. As kg m s^{-2} = N, we have 9.81 10^4 N m^{-2}, a force (newtons) per unit area or pressure. Thus 10 m water column corresponds to a pressure 9.81 10^4 N m^{-2}.

The potential gradient i can be measured simply as $\dfrac{\Delta h}{\ell}$ where Δh is a difference in water column height, and ℓ is the distance between the points of interest. In Fig. 6.9 a simple

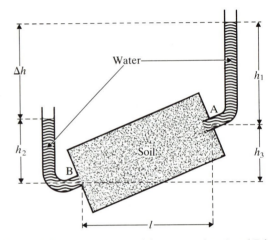

Fig. 6.9 Illustration of potentials at two points A and B in a soil sample. Δh is the 'hydraulic head' determining the flow from A to B. h_1 and h_2 are piezometric levels representing the hydrostatic pressures, and h_3 is a gravitational potential

example is shown where it is desired to measure the rate of flow between A and B. The rate of flow is governed by $\dfrac{\Delta h}{\ell}$. At A and B there are *piezometric levels* of h_1 and h_2 respectively which measure the pressures at those points. The flow of water is obviously aided by the higher position of A relative to B, and h_3 is a component gravitational potential arising on that account.

Consequently the gradient is:

$$\frac{h_1 + h_3 - h_2}{\ell} = \frac{\Delta h}{\ell}$$

Because Δh and ℓ are both measured as lengths $\dfrac{\Delta h}{\ell}$ is dimensionless. The flow equation becomes:

$$q = \frac{k \, \Delta h \, A}{\ell} \qquad [6.6]$$

with the consistent units:

$$m^3 \, s^{-1} = m \, s^{-1} \, m^2$$

that is, k must have the units $m \, s^{-1}$ while the flow is given in $m^3 \, s^{-1}$.

In the example it is assumed that only matric and gravitational potentials are involved. If other potentials are involved it is easy to convert these also into equivalent heights – which may be negative. Note too that the difference in hydrostatic pressure between A and B is merely $h_1 - h_2$, although we also regard h_3 as though it were an increment of pressure. Δh is often called 'hydraulic head'.

This approach proves convenient for many hydrological, geomorphological and geotechnical applications, where elevation differences are common and water columns are frequently a part of measurement devices.

Problems relating to ground water resources, of drainage or irrigation, and not least the movements of ground water in relation to large engineering structures, dams, buildings or other features, commonly involve location of water tables, the assessment of water pressures elsewhere, and the recognition of elevation differences. It is also relatively easy to visualized the meaning of [6.6] when pressure or potential is represented as a height of water. The equation constitutes the well-known Darcy's Law, postulated by him in 1856.

The permeability k has a range of values in saturated soils from about $10^{-12} \, m \, s^{-1}$ for some clays to about $10^{-3} \, m \, s^{-1}$ for coarse sands (Fig. 6.10). The values of k refer to the velocity of the water under a gradient of head, $\frac{\Delta h}{\ell}$ of 1. Permeabilities can be measured in the laboratory using a permeameter. Several types exist, resembling the arrangement in Fig. 6.9 in principle (Lambe 1958), and [6.6] is solved for k. Permeabilities can also be determined by

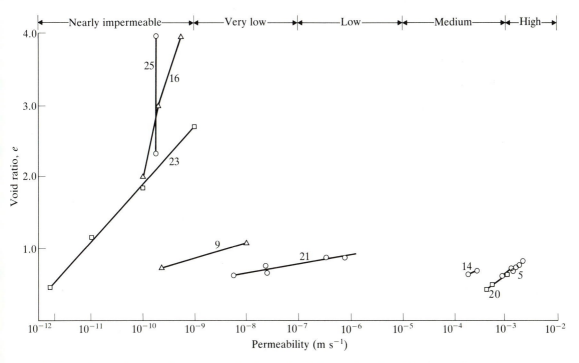

Fig. 6.10 Hydraulic conductivity values for several soils 16, silt, 23, Boston blue clay; 25, montmorillonite clay; 9, silt; 21, silt; 14, 5, and 20, sands. The permeabilities are shown as a function of void ratio. (From Lambe and Whitman 1979)

field observations of flow in wells, or of water loss from pipes under controlled conditions. Values of permeability are not usually attainable with great precision, and the permeability of a saturated material varies somewhat with degree of compaction (Fig. 6.10), and the presence of cracks or other small discontinuities. Quite often permeability varies with direction, being for example, greater along bedding planes than across them.

These variations are usually less significant than those associated with desaturation. As air replaces water in a soil, the paths for the water movement are progressively reduced, and become more tortuous. The permeability of unsaturated soils (or hydraulic conductivity as it is then usually called) falls with water content, so that any statement of permeability must also give the corresponding water content, to have meaning. Normally the permeability is shown as a function of water content (Fig. 6.11) or of matric potential as obtained from a suction-moisture content test. Flow of vapour may also complicate the situation. Permeability often decreases several orders of magnitude in coarser-grained soils in going from zero suction or saturation, to a suction of 10 m water column. Measurement of the permeability of unsaturated soils is fairly difficult although a number of methods have been devised (Staple 1967).

6.13 Distinctions between 'negative pore pressure', 'suction' and 'potential'

At this point it is useful to review certain terms, and clarify our use of them. 'Negative' pressure simply refers to a pressure lower than atmospheric, the latter conventionally being taken as zero. Negative pore pressure refers to a condition of the water in a soil whereby it would be in hydraulic equilibrium with free water at the negative pressure. The soil water does not necessarily have such a negative pressure in an exact physical sense. It behaves substantially however as if it has.

Suction refers to a condition of the soil water dependent upon an interaction of the water and other constitutents of the soil. When the soil is surrounded by gas (e.g. air) at 'atmospheric' pressure, as in a suction-plate, suction is numerically equal to the negative pore pressure as defined. But experience with the pressure plate and pressure membrane

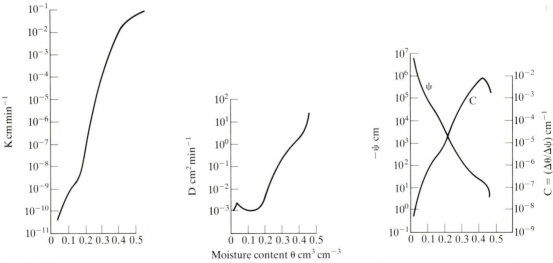

Fig. 6.11 The dependence of hydraulic conductivity, diffusivity and matric potential on volumetric water content for Guelph loam. (After Staple 1967)

apparatus has shown clearly that a pressure in the surrounding gas does *not* significantly change the suction corresponding to a particular moisture content. The water pressure is raised equally, even to a positive value. Thus, suction is then *not* equal to pore pressure. This is important because suction colloquially implies sub-atmospheric pressure, and this was clearly in the minds of earlier investigators. They did not have in mind the pressure of the surrounding gas being other than atmospheric.

Confusion can often be avoided by use of 'potential'. The word lacks the emphasis on 'pressure' and also implicit in 'potential' is the careful definition of a datum. With an appropriate datum, the potential may be numerically equal to the suction, as is the case when suction-plate apparatus is used to determine matric potential, and the datum for the latter is as defined by ISSS (1963). A statement of which, or how many component potentials are involved is preferable in so far as the term 'suction' for what is measured, involves uncertainty as to osmotic effects and may lead to their being overlooked.

The potential concept permits consideration of elevation differences, whereas suction merely refers to a condition induced by the soil material itself. Although suction is measured as the difference between water pressure and gas pressure, with the latter necessarily being greater, it is logical to consider suction as having a negative sign. Without the negative sign water would apparently flow counter to a suction gradient. Potential may have a positive or negative value depending on the datum selected. If water at two points A and B, has potentials (measured relative to the same datum) of –4 and +3 respectively, the driving force in the direction B → A is represented by +7.

While the potential term is favoured in consideration of water movement, the suction term is unobjectionable and falls naturally in certain considerations as those involving soil mechanics, or in usage such as 'suction-water content apparatus'.

Because the fundamental thermodynamic definition of Gibbs free energy explicitly involves temperature, whereas potential does not, free energy will be used when discussing topics involving temperature. Readers delving into the literature of thermodynamics are warned that they will find somewhat similar functions, the Helmholtz free energy, chemical potential, activity etc., certain of which are favoured by workers in other fields. A much fuller knowledge of thermodynamics than has been outlined, is necessary to appreciate the relationships between such terms and those used in this book.

6.14 Changes of moisture content

In Chapter 5, it was pointed out that the flow of heat into a layer is often different to that leaving it. The progressive accumulation or loss of heat energy is associated with change of temperature at a rate depending on the heat capacity. Analogous situations occur with respect to flow of water and changes of moisture content, and can also be analyzed with so-called diffusion equations (although the process is not usually one of molecular diffusion as defined on p. 24).

Moisture content change in soils which remain saturated are necessarily associated with volume change. If we ignore the effect of vibration or similar disturbance leading to reorganization of particles, these circumstances occur only in compressible soils, that is, those that contain clay size particles. Volume decreases of this kind constitute consolidation (Chapter 4) which is extremely important in soils engineering, and in other respects. In 1936 Terzaghi and Frohlich published an equation giving the change of pore pressure u with time t:

$$\frac{du}{dt} = \frac{c_v d^2 u}{dz^2} \qquad [6.7]$$

The analogy with [5.11] is apparent. c_v is a coefficient of consolidation and $c_v = \dfrac{k}{\rho_w m_v}$

where k is permeability, and m_v is the coefficient of volume compressibility, or the compression of the soil (by loss of water) per unit thickness due to a unit increase of pressure. The term $\dfrac{d^2 u}{d z^2}$ is the rate of change of the pore water pressure gradient. When a load (pressure) is applied to compressible soils, the pore pressure immediately rises (establishing pressure gradients) and this increment of pressure then dissipates slowly as water leaves the stressed zone. Consolidation is complete when u has returned to its original value (or the value determined by the hydrological environment). The equation can be used to establish the degree of consolidation (and the water content) occurring with time. The equation is described in detail in for example, Terzaghi and Peck (1967).

In unsaturated soils questions of flow of water necessarily involve more complex equations than [6.5]. A gradient of matric potential is associated with a marked moisture content gradient. This follows from the nature of the suction (i.e. matric potential) – water content relationship. It implies a substantial variation in conductivity, and consequently variations in flow, along the gradient. Flow is therefore inevitably associated with significant, concurrent moisture content changes on this account. This is in contrast to the situation in saturated soils where k is usually only slightly dependent on u (or, more precisely on the effective stress $\sigma - u$). Consequently in uniform saturated soils where there is a linear gradient $\dfrac{d u}{d z}$, a steady state flow, $q = k\,i\,A$ ([6.5]) can usually be assumed. In unsaturated soils (i.e. with air occupying some pore space) a steady state flow is only possible under a certain, non-linear, gradient, and is probably rare. A linear gradient is short-lived because of the simultaneously changing moisture content in the z direction, and the associated changing gradient $\dfrac{d^2 u}{dz^2}$.

The flow q_u at a specified position in an unsaturated soil, can be expressed by (Hillel 1971):

$$q_u = k_u\,\frac{\partial \psi}{\partial \theta}\,\frac{\partial \theta}{\partial z} \qquad [6.8]$$

where k_u = hydraulic conductivity
  ψ = potential (matric)
  θ = moisture content
  z = distance

The term $k_u\,\dfrac{\partial \psi}{\partial \theta}$ is called the soil water diffusivity. It is of course dependent on moisture content, and values can be obtained experimentally (Fig. 6.11). If k is in m s^{-1}, and ψ is expressed as m of water column, then the diffusivity is seen to be m^2 m^{-1}.

In [6.8] the gradient of moisture content $\dfrac{\partial \theta}{\partial z}$ is, in effect, used as a measure of the gradient of potential. The latter gradient is represented by the product of $\dfrac{\partial \theta}{\partial z}$ and $\dfrac{\partial \psi}{\partial \theta}$ (from the diffusivity term). Looked at in this way the equation is similar in form to [6.6]. However, it is often more convenient to work with moisture contents and diffusivity values, than with permeabilities of unsaturated materials and potential gradients.

Equations derived from [6.8] can be used for determination of moisture content changes, when the variation $\dfrac{d q}{d t}$ of flow with time, or distance, $\dfrac{d q}{d z}$, is known, together with the suction-moisture content relationship of the material which gives $\dfrac{d \psi}{d \theta}$. The equations are in several respects similar to those used in studies of heat flow and temperature.

Questions of moisture movement in unsaturated soils are of importance particularly in agronomy, and there has been much work on, for example, infiltration of water from the ground surface, and the subsequent distribution of moisture content, and potential, with time and depth (see Chapter 10). The computations involved are fairly complex and the reader is referred to textbooks such as Baver, Gardner and Gardner (1972).

7

Water vapour and ice in soils

7.1 The significance of the three phases of water

Water occurs abundantly at the earth's surface as vapour and as solid, as well as in the liquid phase. The hydrological cycle between ground and atmosphere with its attendant dissipation of energy and its land forming effects, and the widespread effects of snow and ice, of freezing and thawing, arise because of this circumstance.

Although below its boiling point, water at 'normal' temperatures has a high vapour pressure compared to most earth materials. Thus a highly significant exchange of molecules between water and air occurs. As the vapour holding capacity of the air changes with temperature, there are frequent and very significant fluxes of energy and mass, associated with evaporation and condensation.

Energy exchange involves the latent heat of fusion of ice, exactly as the latent heat of evaporation is involved in vaporization or condensation. The presence of ice radically changes the mechanical properties of frozen soil; but the properties associated with the frozen state are made more complex by the common occurrence of water, along with the ice, at temperatures below the 'normal' freezing point of 0 °C.

Phase transitions are not limited to a single temperature, whether it be boiling 'point' or freezing 'point', under the complex conditions associated with the materials of the earth's surface. Water is an integral part of soils: transitions between the phases of water are influenced by properties of the solid constituents. In turn, phase transitions greatly affect the behaviour of the soil as a whole. The natural consequences are everywhere evident in association with drying, freezing, or thawing of soils. The consequences for man's construction activities are particularly immediate in cold regions. The geotechnical engineer is then working with a material close to or indeed at its melting point. This is a situation usually quite unthinkable to the structural or aeronautical engineer whose solid working materials are at temperatures hundreds of degrees below their melting point.

7.2 Vapour pressure of soil water; relation to potential

'Vapour pressure' appears frequently in climatological studies. Not least for this reason it is important to understand vapour pressure as it relates to soil water. Consider some water placed in a sealed container, which it does not completely fill. Molecules constantly leave and enter through the water surface in an apparently random fashion, except that until there is a certain uniform concentration of vapour in the space (which may also contain air) there is a net loss from the water. This critical concentration is defined by the *saturation vapour pressure*, that is, the pressure exerted by that concentration. The saturation vapour pressure is the pressure of vapour which is in equilibrium with the water (that is, the pressure tends neither to decrease or increase). Apparently random molecular movements, that is, diffusion, equalize the concentration of vapour molecules.

The saturation vapour pressure, increases with temperature (Fig. 7.1). Note that the vapour pressure is distinct from, and except at boiling point less than, that of the air in which the vapour happens to be. The adjective 'partial' is sometimes used to stress this, as in 'partial pressure of vapour'. The sum of the vapour pressure and the air pressure is the total pressure of the moist air.

The pressure of vapour in equilibrium with soil water is commonly less than that in equilibrium with pure, free water. The pressure varies with the free energy or potential of the soil water. Only when the potential of the soil water is zero (when the soil water would be in thermodynamic equilibrium with an adjacent body of pure, free water) is the equilibrium vapour pressure of the soil water equal to that of pure, free water, that is, to the saturation vapour pressure.

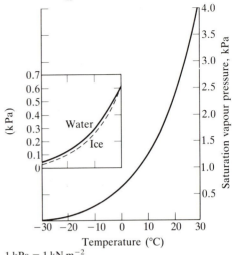

$1\,kPa = 1\,kN\,m^{-2}$

Fig. 7.1 Dependence of saturation vapour pressure over plane surface of water, on temperature. At 100 °C the vapour pressure approximates the pressure of the atmosphere ($\simeq 100\,k$ Pa) Inset: Saturation vapour pressure over water and ice at temperature below 0 °C. (After Byers 1965)

Air containing less vapour than that giving the saturation vapour pressure is said to have a *relative humidity* given by:

$$\text{R.H.} = \frac{\text{actual vapour pressure, } P}{\text{Saturation vapour pressure for pure free water, } P_0}$$

Soil water in equilibrium with vapour of pressure P in the air, is said to have the same relative humidity. Thus we speak of the *relative humidity of the soil water*, which is:

$$\frac{\text{pressure of vapour in equilibrium with soil water, } P}{\text{saturation vapour pressure for pure free water, } P_0}$$

Sometimes, this is loosely referred to as the relative humidity of the soil. In defining relative humidity of the soil water, P_0 refers to pure free water at the same temperature as the soil.

If the relative humidity of soil water and the relative humidity of the surrounding air are not equal, water evaporates or condenses upon the soil. The magnitude of the difference between the relative humidities influences the rate of transfer between soil and air, although other factors such as mixing due to movement of the air are also very important. If the air is saturated (i.e. saturation vapour pressure exists) then water is transferred to the soil unless the potential of the soil water is zero. If the evaporating surface is *not* at the same temperature as the air, evaporation or condensation occurs even if the relative humidities of the soil water and the air are equal. The relative humidity of soil water does not change greatly with temperature (Coleman 1949). But vapour pressures increase rapidly with temperature, as indicated in Figs. 7.1 and 7.2. Soil water of relative humidity 60 per cent at 20 °C will have a vapour pressure of 1.403 kN m^{-2} (= 10.5 mm Hg). If the air has a relative humidity of 60 per cent but is at a temperature of only 10 °C, its vapour pressure

is only 0.737 kN m^{-2} (= 5.53 mm Hg). Consequently, evaporation from the soil will occur.

The amount of vapour in the air, can also be represented by the concentration of vapour molecules expressed, for example, as the mass of vapour in unit volume of the air. This measure is the *absolute humidity*, and because the pressure depends on the concentration, appropriate calculations also permit relative humidities to be found from the ratio of the absolute humidities.

There is a fundamental relationship between vapour pressure and potential (or free energy). This illustrates again the power of the potential concept. As potential falls with decrease of soil moisture content the equilibrium vapour pressure falls. It is well known that the vapour pressure of a solution is less than that of pure

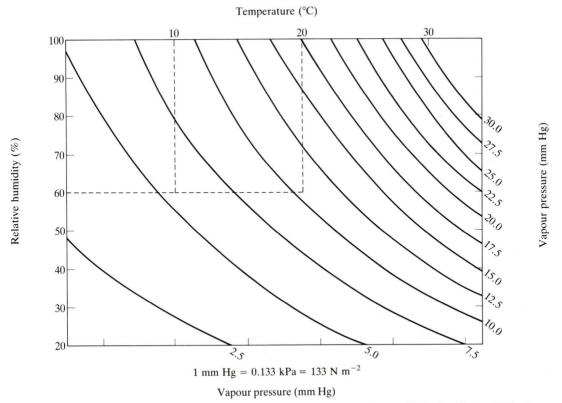

1 mm Hg = 0.133 kPa = 133 N m^{-2}

Vapour pressure (mm Hg)

Fig. 7.2 Relationship between temperature, relative humidity and vapour pressure (modified after Mather 1959). A temperature of 10 °C with 60 per cent relative humidity gives a vapour pressure of 5.5 mm Hg (= 733 N m^{-2}). With 20 ° and 60 per cent relative humidity, the vapour pressure is 10.5 mm Hg (= 1400 N m^{-2}). The intersection of the curves with the temperature scale (that is, 100 per cent relative humidity) shows saturation vapour pressures – compare Fig. 7.1

water. Adsorption forces have a similar effect on the soil water, and so do capillary forces. These causes for the lowered vapour pressure are the same as those for the lowered potential.

In accordance with the fundamental definition of potential, water vapour is in equilibrium with adjacent water in a soil when the potentials of the two phases are equal:

$$\psi_{sw} = \psi_{vap}$$

Because the water vapour is pure (although the soil water is not) its potential is dictated by its pressure alone; the vapour pressure expressed as relative humidity is therefore a direct measure of the soil water potential. When the soil water has several component potentials, the relative humidity is a convenient measure of the total potential.

Figure 7.3 shows the relation of relative humidity to soil water potential, this being expressed as cm water column. (1 cm water column $\simeq 10^2$ N m^{-2}). Comparison of this

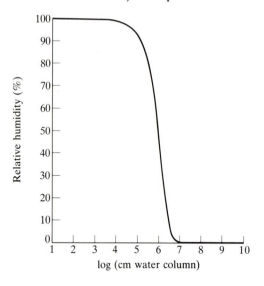

Fig. 7.3 Relationship between potential and relative humidity at 20 °C. Note that 100 cm. of water column corresponds to a pressure at 9.8 10^3 N m^{-2} and a free energy difference of 9.8 J kg^{-1}.

figure with Fig. 6.3, shows that for damp or wet soils the relative humidity is high (that is, the equilibrium vapour pressure is near that of pure water) but in soils that are 'dry' in an

everyday sense, the relative humidity falls rapidly with decrease of moisture content. The relationship of water potential to relative humidity can be demonstrated theoretically as follows. Consider a model which consists merely of a vertical column of vapour, of unit cross section. Soil water in equilibrium with the vapour at any level would, as noted, have the same potential. The pressure difference dP_{vap} between two points vertical distance dZ apart, in the column is:

$$dP_{vap} = \rho_{vap} \, dZg \qquad [7.1]$$

There is a change of potential dψ per unit mass, corresponding to the change dP_{vap}. This is given by the work involved in moving unit mass of vapour through dZ, and is therefore dZ g (units kg m^2 sec^{-2}). The definition of potential in terms of work was given on p. 23. To *move unit volume* involves*: $\rho_{vap} \, dZg$ which is however, as noted, the pressure difference over distance dZ. Further,

$$\rho_{vap} = \frac{nM}{V} \quad \text{where } V \text{ is any volume; } n = \text{number}$$

of molecules in volume V; and $M =$ molecular weight. So from [7.1]:

$$\frac{dP_{vap}}{dZ} = g\frac{nM}{V} \qquad [7.2]$$

According to the gas law:

$$PV = nRT \qquad [7.3]$$

where $P =$ pressure
$V =$ volume
$n =$ number of molecules
$R =$ universal gas constant J mole^{-1} K^{-1}
$T =$ absolute temperature K

in [7.2] V can therefore be replaced by: $\dfrac{nRT}{P_{vap}}$ to give: $\dfrac{dP_{vap}}{dZ} = \dfrac{gMP_{vap}}{RT}$

and rearranging: $dZ \, g = \dfrac{RT}{M} \dfrac{dP_{vap}}{P_{vap}}$

* This expression is not formally correct in that the units of $\rho_{vap} \, dZ \, g$ (kg m^{-3} m m s^{-2} = kg m^{-1} s^{-2}) are not those of work or potential (kg m^2 s^{-2}) – but the ρ_{vap} is to be regarded as merely a statement of the mass in our defined volume of vapour (which happens to be unit volume, m^3), and the units are then indeed kg m m s^{-2} = kg m^2 s^{-2}.

but dZ g is the change of potential per unit mass, dψ so:

$$d\psi = \frac{RT}{M}\frac{dP_{vap}}{P_{vap}} \quad\quad [7.4]$$

Now, assume $P_{vap} = P_0$, the saturation vapour pressure of a datum body of pure water, and dP_{vap} equals a differential increment to the vapour pressure of soil water P. Then, for a finite value of ψ integration of [7.4] gives:

$$\psi = \frac{RT}{M} \ell n \frac{P}{P_0}$$

$\frac{P}{P_0}$ is the relative humidity of the soil water. This derivation is discussed in Rose (1966). ψ has the units J kg^{-1}, and the relation is illustrated in Fig. 7.3, where ψ is represented as cm of water column.

The relationship [7.5] (Fig. 7.3), is utilized in methods for determining suction of soil water, by measurement of the pressure of vapour in equilibrium with soil water (the 'psychometric' technique). Alternatively, the moisture content of a sample can be measured after allowing it to come to equilibrium with controlled vapour pressures (the 'vapour pressure method' for determining suction-moisture content relationships, Croney, Coleman and Bridge 1952). The procedures give suction values that unequivocally include both the matric and osmotic components. Because the vapour pressure cannot be measured or controlled sufficiently accurately the methods cannot be used for situations involving small suctions. The vapour pressure then changes only very slowly (cf. Fig. 7.3) with suction, and the use of the vapour pressure method becomes valuable only where suctions exceed about 3 10^2 m water column. It is then extremely useful as it obviates the need for very high pressure apparatus of the pressure membrane type.

In the foregoing discussion no reference has been made to gravitational potential (that due to elevation difference) in relation to the vapour pressure of the soil. Consider a column of vapour (in air) again, in equilibrium with a body of free water at the base Fig. 7.4. There is a sample of moist soil adjacent to the water. The vapour pressure of the soil water is P, and the relative humidity for the soil water is P_0 where P_0 is the saturation vapour pressure of the free water. If the sample is raised to some elevation above the free water, the soil water will have the same vapour pressure, P. The vapour pressure P_0 in the column is necessarily lower at a height above the free water, and accordingly $\frac{P}{P_0}$, the relative humidity (and potential) of the soil water has been increased. A component potential has been established, by virtue of the elevation difference between the soil water and the 'datum' – the level of the free water.

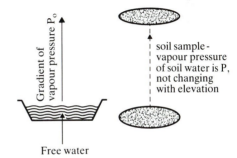

Fig. 7.4 Diagram to illustrate that the relative humidity of soil water ($\frac{P}{P_0}$), and therefore potential, is increased by elevation. (see text)

Elevation differences that may be highly significant in other respects, are nevertheless associated with very small changes in relative humidity. Consequently, vapour pressure considerations have little application in questions involving elevation differences of soil water.

7.3 Movement of water vapour in soils

Movements of air as caused by wind, carry with them the molecules of vapour present. This bulk movement of air does not normally occur within the soil with uniform temperatures, except on a very minor scale in response to local inequalities of air pressure.

The vapour pressure in the soil pores varies, as we have seen, with the potential of the soil water. Therefore gradients of soil water potential represent gradients of vapour pressure. Equally these constitute gradients of concentration of vapour molecules and give rise to movement by diffusion. Diffusion involves molecules moving in all directions, but results in a net movement, a flux in the direction of a gradient of vapour density given by:

$$Q_{vap} = D_o \, \partial \, \rho_{vap} \text{ kg m}^{-2} \text{ s}^{-1} \qquad [7.6]$$

The partial differential (symbol ∂) is used because D_o (m^2 sec^{-1}), the molecular diffusion coefficient of water vapour through air, is not a constant. It increases with temperature and decreases with total pressure, and partial pressure of the vapour, and thus with ρ_{vap}.

Because of the tortuous path, with obstructions of particles and liquid water, vapour diffusion is slower in soils than in free air. In wet soils, where the gas-filled passages are not continuous, that is, where the suction is less than the air intrusion value (see p. 76), [7.6] is barely relevant. Where there are isolated, gas-filled pores, vapour may cross such a pore and condense. This can be followed by evaporation of water into another gas-filled pore further along the potential gradient. Such a process is likely to be slower than diffusion through unsaturated drier soils. The proportion of water movement in the vapour phase relative to that in the liquid phase, increases from zero (at saturation, without gas inclusions) to a value of more than 1 : 1 (for soils with substantially vapour-filled void space).

The concept of potential can only be applied to isothermal situations. Vapour pressure and density are of course dependent on temperature, and temperature differences also result in vapour movement in the soil. The amount of such movement may exceed that due to a gradient of potential if such also occurs.

Overall, rather little has been firmly established about vapour movement in soils. Its direct practical importance, for example in an agricultural context, remains uncertain. For unsaturated soils, vapour movement compounds the difficulties associated with evaluation of soil thermal conductivity.

7.4 Composition of frozen soils

Frozen ground at natural temperatures has many special properties (Anderson and Morgenstern 1973). There is a tendency to regard a frozen soil simply as one in which the water has been replaced by ice. This involves an error relating to a fundamental characteristic of frozen soils: at most temperatures of interest frozen soils contain ice *and* water. In many clays as much as half the moisture is in the liquid state at temperatures several degrees below the freezing point of pure water.

This fact was first revealed clearly by Nersesova (1953) who found, using a calorimeter, that less heat was necessary to thaw frozen soils, than was required to melt the amount of ice believed present. The discrepancy was due to some of the moisture content being already water. Fig. 7.5 shows

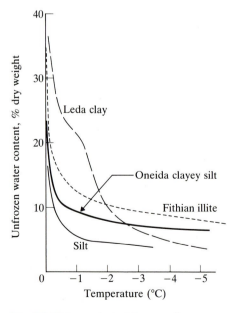

Fig. 7.5 Water content of frozen soils

examples of the amount of liquid water at different temperatures. Similar values have also been determined by dilatometry, with measurement of the volume changes which follow from the well-known 9% contraction on the transition of ice to water (Williams 1976).

The water contents of frozen soils are stable in a thermodynamic sense, remaining unchanged until there is a change, of temperature or pressure. In general, the more finely-grained (or small-pored) a soil is, the greater is the amount of water at a given temperature. The reasons for the presence of the water are capillarity, adsorption and solution – those phenomena which give rise to matric and osmotic potentials. Other sources of potential considered earlier, also influence the amount of water present but in a less direct fashion. Because we are necessarily dealing with temperature, we shall later in this chapter refer to Gibbs free energy rather than potential. Nevertheless, where moisture movement is concerned this can usually be understood as potential in the sense used in Chapter 6.

7.5 Relationship of unfrozen water content to temperature, soil type and suction

Curves of the type in Fig. 7.5 have been compared by Koopmans and Miller (1966) and Williams (1967 and 1976) with the suction-water content relationships (Chapter 6) of the same soils. It was found that the *water* content (Fig. 7.5) at a particular temperature, while varying from soil to soil corresponded to the same suction value. That is to say, reading the water content into the suction-water content relationship for each soil always gave similar suction values. Repeating the process for different temperatures, enables a graph relating freezing temperature and suction to be drawn (Fig. 7.6).

The two sets of experiments that lead to the correlation shown in Fig. 7.6 are at first sight quite unconnected. The suction-moisture content test involves determination, at room temperature, of the decrease of water content, associated with increasing suction. The other set of determinations involves the water content decreasing because the temperature is lowered and water is changed to ice. Since progressively lower temperatures are required to freeze additional increments of soil water, we can say that the freezing point of soil water falls as the *water* content is reduced. Figure 7.6 correlates freezing 'points' and suctions associated with such water contents.

Two conclusions may be drawn which explain the correlation in Fig. 7.6. Firstly, as the water content is reduced by progressive formation of ice, the remaining water is indeed under an increasing suction. Just as when water is transferred to vapour or is drained from the soil so also when water is transferred to ice the remaining water experiences suction. Secondly, the occurrence of suction reveals the cause of the depression of the freezing 'points'. In fact this latter conclusion was early drawn by Schofield and DaCosta (1938), although on the basis of a different experiment: they observed that freezing commenced at temperatures less than 0 °C in soils the water contents of which were less than saturation (and thus had suction). Schofield (1935) pointed out that the suction, in fact, represented the free energy of the soil water, and, as was well known, freezing points depend on the free energy. The relation of suction to freezing temperature is considered from a thermodynamic point of view in section 7.11.

The *total* moisture content (ice and water) does not influence the water content significantly. More ice takes up more volume in ice layers or lenses, but does not affect the disposition of the water. Reducing the total moisture content of the materials illustrated in Fig. 7.5 would merely truncate the upper parts of the curves.

According to Fig. 7.6 cooling a soil to only a degree or two below 0 °C, develops a large suction. The large suctions can be more directly demonstrated. Suctions are identified with

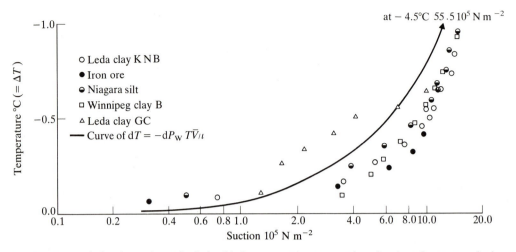

Fig. 7.6 Theoretical and experimental relationship between temperature and suction (sample at atmospheric pressure).

effective stress (Chapter 4). The suctions developed by freezing are revealed by the simple observation that a previously unfrozen clay sample, after being frozen and then thawed, shows a much changed structure. It consists of hard almost shaly, flakes. The clay-flakes have consolidated due to the effective stress associated with the suction. Examination of the sample while still frozen reveals thin layers of ice, which are composed of water drawn out of the intervening layers of mineral soil. On thawing, the ice becomes free water which tends to run out of the sample. Subsequently, the sample exhibits a high preconsolidation pressure. Although normally preconsolidation implies increased strength the discontinuities between the flakes are planes of low strength, and the strength in bulk is therefore not high. These changes of soil structure are important in several respects, not least because they confuse interpretation of the soil with respect to preconsolidation and its relation to other geological and environmental factors.

7.6 Frost heave

A feature of widespread importance is the fact that suctions developed in the ground on freezing lead to movements of water towards the freezing soil. On entering the frozen zone the migrated water becomes ice. The frozen soil now has more ice, a higher moisture content, than before freezing. Consequently the volume of the soil increases, and this constitutes *frost heave*. The additional moisture in a soil showing substantial frost heave is revealed by abundant visible ice in lenses or layers (Pl. 7.1), and by the presence of much free water on thawing.

Frost heave must be distinguished from the 9 per cent expansion that occurs on freezing of water. It is instead the result of *additional* water (as ice) relative to the prefreezing condition. Frost heave is most prevalent in silty soils – field experience has shown it to be greater in regions with a high water table, and where overburden pressures (weight of overlying material) are small. It does not occur in materials that are dominantly coarse-grained. Indeed in the absence of the heave effect, water may be pushed away from the freezing front by the 9 per cent expansion due to ice formation. Although clay-rich materials may show some frost heaving, the low permeability of the unfrozen soil slows the water migration. Frost heaves involving more than 40 per cent volume increase are fairly common in nature and produce local elevations of the ground surface, for example

Plate 7.1 Frost-heaved clay. The dark layers are more or less pure ice. (Courtesy National Research Council, Canada)

during annual freezing, of a corresponding percentage of the thickness of the frozen layer.

The extreme practical importance of frost heave lies largely in the great loss of strength occurring when frost heaved soils thaw – often producing mud. Clearly roads so affected have little bearing capacity, and, in permafrost regions, engineering activities may produce thawing of the permafrost with similar disastrous effects on foundations. But the expansions at freezing, that is the heave itself, are also important for the damage they may inflict on foundations or other buried structures. Large quantities of coarse-grained materials are used in foundations, especially for roads, in cold climates. In this way deleterious heaving of the road surface during the winter is avoided.

Many studies have been made of factors influencing frost heave, and Beskow's (1935) work remains a classic. One of the major difficulties is determining exactly where the suction originates that causes the migration of water to the 'frost line' (the boundary between frozen and unfrozen soil). As there is normally a gradient of temperature through the frost line and into the already frozen soil (Fig. 10.2a,), there is a range of suction associated with the range of temperature involved. It was earlier assumed that as ice substantially blocks the passage of water into the frozen soil, ice accumulation was limited to the frost line, that is, the boundary between frozen and unfrozen soil. Even so, it was not easy to define for a given soil the magnitude of the 'pulling forces' or suction that would be effective in establishing the potential gradient in adjacent unfrozen ground. While sands exhibit little or no such gradient towards the frost line, for silts and clays different values may be measured depending on the temperature gradients, rate of freezing and other conditions. Typical values of suctions measured just below the frost line in different soils are given in Table 7.1.

7.7 Permeability of frozen soils and frost heave

The reason for the uncertainties described above lies in the permeability of the frozen soils. When frozen, soils show greatly reduced permeability which is to be expected following from the transformation of water to ice. However, several investigations (Miller 1970, Burt and Williams 1976) have established that frozen soils are to some extent permeable; this must be partly due to flow of unfrozen water in the frozen soil. The permeability in this

Table 7.1 Typical values of suction measured below an advancing frost line.

Material	Suction $N\ m^{-2}$
Natural soils:	
Silt	$5.5\ 10^3$
Silt	$7.3\ 10^3$
Leda clay (chlorite-illite)	$4.08\ 10^5$
Graded fractions, sample particle diam., 10^{-4} cm	
73–75	$2.6–4.3\ 10^3$
49–73	$1.34\ 10^4$
23–49	$0.7–1.4\ 10^4$
6.4–23	$2.5–4.0\ 10^4$
1.7–6.4	$8.2\ 10^4$
0–75	$6.5\ 10^4$

From Williams (1967).
Note: Under slow and prolonged freezing the values would tend to increase (see text).

instance would be similar to that for the soil in the partly saturated state, the presence of the ice being immaterial in this case. It is the amount and distribution of the water that would count, the volume of water dictating the cross-sectional area of the actual flow paths. In

Fig. 7.7 permeabilities of various frozen soils are shown as a function of below 0 °C. There is clearly some correlation with unfrozen water contents (Fig. 7.5). Although the permeabilities are much lower than for the same materials at above-freezing temperatures they are by no means negligible. However, recent work suggests that there is also an important transfer in the ice phase, this being associated with regelation – a freezing at the 'upstream' edge of an ice inclusion and thawing on the downstream edge (Horiguchi and Miller 1979). This may be important at temperatures below that at which there is abundant water in the frozen soil, or in relatively coarse-grained frozen soil where unfrozen water contents are always small. It appears that a layer of ice transverse to the direction of flow does not represent the barrier to water flow that would otherwise be expected (Burt and Williams 1976).

If the advance of the frost line is slow (that is if freezing is slow) then, particularly in fine-grained materials, a significant quantity of water may move slowly into a layer of already frozen soil. It does so under the influence of the larger suctions (lower potentials) developed *within* the frozen soil where temperatures are lower. Such movements have been demonstrated experimentally; they give rise to *secondary* frost heave. If a long enough time is available, some such migration, perhaps of vapour, can also be detected in coarse-grained materials not normally susceptible to heaving.

When freezing is more rapid, and when the permeability of the frozen ground is low, no significant quantity of water penetrates the frozen layer. The suctions effective in causing the water migration to the frost line are then only those developed in association with ice formation very close to the frost line, where temperatures are near 0 °C and the suctions correspondingly small. But this does not rule out the possibility that over perhaps years, there may be gradual ice accumulation within the frozen ground. It should be noted that the temperature at the frost line is somewhat lower in fine-grained, and therefore fine-pored, soils.

Consequently the suction at the frost line will be greater in such soils as indicated in Table 7.1. The significance of the pore size in this connection is discussed in the next suction.

7.8 Freezing and drying compared further: the role of pore size

The association of frost heave with suction, draws attention to other similarities between frozen soils (in which part of the soil water is replaced by ice) and unsaturated soils (in which part of the soil water is replaced by air). In so far as suction in the latter is due to capillarity, similar relations apply in frozen soil.

The equation:

$$P_a - P_w = 2\frac{S_{aw}}{r}$$

where S_{aw} = the surface tension, air-water.
P_a = pressure of air
P_w = pressure of water
r = radius of meniscus $\triangleq$ radius of containing pore

which is [6.2] (without the term $\cos\theta \triangleq 1$), is replaced by:

$$P_i - P_w = 2\frac{S_{iw}}{r} \qquad [7.7]$$

in which S_{aw} = the surface tension, ice-water
P_i = pressure of ice

Although it is not clearly established that concave ice-water menisci (radius $\triangleq r$) develop in the pores (and the crystal properties of ice suggest a more complex interface) the relationship apparently holds true at least for temperatures near 0 °C. At lower temperatures the water content is sufficiently low that water occurs only in layers around the particles.

Surface tension, as [7.7] indicates, is not a phenomenon limited to gas-liquid interfaces. It is associated with any interface between materials or phases, although often referred to as interfacial energy (units J m^{-2}). The value of surface tension for ice-water interfaces is about 30 mN m^{-1} compared with 72.75 mN m^{-1} for

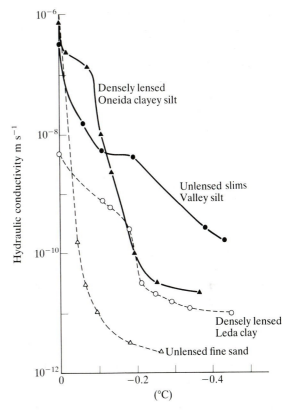

Fig. 7.7 Hydraulic conductivity of frozen soils at temperatures near 0 °C

air-water at 0 °C. Consequently it might be expected that the suction for a given water content would be greater in the unsaturated (unfrozen) soil than in the frozen soil by a factor of $\frac{72.75}{30}$. There is evidence that this is the case, and the data used in Fig. 7.6 have been so adjusted.

There are exceptions, however. Air does not occupy pores within a soil unless a suction greater than the air intrusion value (p. 76) has been developed. Ice does not spread into a soil until an 'ice-intrusion value', also a particular suction value, related to pore radii in a complex manner, is reached. Prior to this the ice must be restricted to some larger opening, a crack perhaps, or to the soil surface. The suction may then be exactly what it would be were the same moisture content reached by drying, or drainage, to the same water content

(Koopmans and Miller 1966). The ice-intrusion value, of course, constitutes the suction at a penetrating frost line, at least during relatively rapid freezing, and is greater the more fine-grained the soil.

As ice increases and water decreases in the soil pores, the significance of adsorption forces on the water becomes greater. The same applies on reducing the water content of unfrozen soils. Osmotic effects due to dissolved salts are relevant to consideration of suction and free energy in unfrozen soils, and obviously, dissolved salts have an effect on the freezing 'points' of soil water just as they do for water in general.

All these circumstances suggest further examination of the Gibbs free energy, which (exactly as potential) can be regarded as measuring the combined effects of several components (component potentials). This is considered in section 7.11.

7.9 Pressures in freezing soils: basic equations

In Chapter 6 it was found useful to consider potential or suction, as due solely to pressure in the normal macroscopic sense. A similar approach is useful in the case of frozen soils, as has already been demonstrated in the previous sections. The ice in frozen soils, in so far as it occurs in bodies larger than pore size is in fact more or less pure bulk ice. The pressure on the ice is simply that applying to the soil itself. If the soil is at depth, then the ice will be carrying the overburden pressure. Ice which extends into the pores, because it forms by progressive extension of the larger ice masses described, is in thermodynamic equilibrium with it. Increments of ice cannot be formed in such a way as to lead to non-equilibrium situations. Consequently, we can assume that ice in the pores has the same pressure as the bulk ice, even though we know that the condition of the ice in pores in a

precise sense is not necessarily solely due to 'pressure' as normally understood.

We can similarly regard the thermodynamic condition of the water in the frozen soil in terms of an equivalent pressure. However, that pressure of the water is not, as amply demonstrated, equal to the pressure of the ice. If it were, it would be impossible to explain the coexistence of the ice and water at temperatures well below the normal freezing point of 0 °C.

A relatively simple thermodynamic analysis (Edlefsen and Anderson 1943) based on the definitive equation for dG the relative free energy (see p. 24) shows that the freezing temperature (or freezing 'point') is related to $P_i - P_w$ according to:

$$dT = - dP_w \frac{T\bar{V}}{\ell} \qquad [7.8]$$

where dT = lowering of freezing point °C
$-dP_w$ = pressure on water relative to that on ice
T = absolute temperature K
$\bar{V}$ = specific volume of water
ℓ = latent heat of fusion

Equation [7.8] shows that the temperature at which ice and water exist in equilibrium, falls by 0.081 °C for each 1 10^5 N m^{-2} reduction of pressure on the water *only*. This equation is illustrated in Fig. 7.6. dP_w refers to the difference in pressure $P_i - P_w$, and is of course, the suction. We can say alternatively that lowering the temperature by 1 °C induces a suction of 1.2 10^6 N m^{-2}.

This, and a further equation, are extremely important in understanding the behaviour of freezing soils. If pressure dP is applied to the two phases, the change of freezing point dT is:

$$\frac{dP\,(\bar{V}_w - V_i)T}{\ell} \qquad [7.9]$$

where $\bar{V}_i$ = specific volume of ice

In this case, a change of freezing point of 1 °C involves 1.33 10^7 N m^{-2}, or 0.0074 °C per 1 10^5 N m^{-2} addition of pressure (to both phases). Equation 7.9 may be recognized as that for the effect of pressure on freezing point normally given in elementary physics texts. As we have seen, the situation of water and ice in freezing soils is not entirely 'normal' in this respect.

The two equations are sufficient to enable us to quantify the pressure relationships of freezing soils in general. Suppose that the pressure of the ice can be taken as atmospheric, as is the case, for example, in the freezing of saturated soils near the ground surface. As freezing progresses and the temperature falls, negative pore pressures develop according to the temperature as shown in Fig. 7.6 or [7.8]. The negative pore pressures are equal to the suction. Depending on the pore pressure (we can equally well say potential) in the underlying unfrozen soil, a potential gradient may or may not be set up such that a flow (in the unfrozen ground) towards and into the freezing layer occurs. As noted in section 7.7 exactly which temperature of the freezing layer, and thus which pressure dP_w in the freezing soil will define the gradient is rather uncertain.

When frozen ground occurs at depth the pressure of the ice is greater than atmospheric by the overburden pressure (the total stress σ – Chapter 4). The negative pore pressure dP_w is less (the pore pressure is raised in an absolute sense) by an equal amount. The suction, the difference between the ice and water pressures, remains the same. Under sufficiently high overburden pressure the pore water pressure may even be positive. The higher pore water pressure (or 'less negative') means that the potential gradient towards and into the freezing soil is likely to be reduced. This is the reason for the observation that overburden pressure reduces frost heave.

When the ice pressure is raised above atmospheric in this manner the freezing temperature-pressure relationships are described by [7.8] and [7.9] together. Application of [7.9] accounts for the increment of pressure in both phases due to the overburden pressure, and [7.8] accounts for the lower pressure of the soil water relative to that of the ice. The combined equation is:

$$dT = \frac{dP(\bar{V}_w - \bar{V}_i) - (dP_i - dP_w)T\bar{V}_w}{\ell} \qquad [7.10]$$

where $dP = dP_i$ = overburden pressure, pressure of ice

$$dT = -\,°C$$

or: $dT = \dfrac{(dP_w \bar{V}_w - dP \bar{V}_i)T}{\ell}$

This equation predicts that the effect of overburden pressure is to displace slightly, curves of the type in Fig. 7.5 towards lower temperatures. Such displacement has been observed experimentally (Williams 1976).

When water freezes in a closed space very large expansive pressures are produced in accordance with [7.9]. Thus cooling to say, $-0.5\ °C$ produces $6.5\ 10^6$ N m^{-2}. However, no additional water moves into the space and the maximum possible expansion is the approximately 9 per cent associated with the transformation of water to ice. If such expansion is possible without rupture of enclosing rock nothing further happens. If the rock does not yield then the pressures are those given by [7.9], according to the lowest temperature reached. Such high pressures are of course, frequently more than sufficient to shatter the rock. Other aspects of such situations were considered briefly in Chapter 2

The pressures developed by frost heave are quite different. The volume of ice is augmented by the arrival of water and this situation must be clearly distinguished from that described in the previous paragraph. The maximum pressures generated by frost heave are much less. As soon as water ceases to be added, there is no further increase in the frost heave pressure. The additional water forms lenses or other ice masses larger than pore size, thus producing the heave. The heaving pressure is simply the pressure of the ice. It represents the ability to lift the overburden or overcome similar constraints. The heaving pressure depends on the temperature of the particular piece of soil under consideration as described by [7.10]. Normally however, the temperature is not the limiting factor. It is the availability of water for enlargement of the ice which is decisive.

Suppose for example, that the water in the unfrozen soil immediately below the frost line happens to be at atmospheric pressure. The pressure of the water in the freezing soil falls, to provide, together with the ice pressure, the suction $P_i - P_w$, which is dictated by temperature. Water will migrate into the frozen soil and will continue to do so, in principle, until the water in the frozen soil again has atmospheric pressure. As water migrates and ice forms, if there is some constraint (overburden etc.) there will be a corresponding pressure P_i on the lenses or other ice bodies. The maximum value of P_i is limited because P_w cannot rise above atmospheric pressure; at that maximum there is no longer any gradient to cause water to move to the freezing zone. We see that the maximum frost heave pressure depends on the pore water pressure P'_w of water available for eventual migration to the freezing material. Thus the maximum frost heave pressure P'_i is simply:

$$P'_i = P'_w + (P_i - P_w)_T \qquad [7.11]$$

where $(P_i - P_w)_T$ is the suction, which of course is determined by the temperature, T. Unfortunately we rarely know what temperature is relevant. If water is unable to penetrate into frozen soil because of its low permeability, as discussed in section 7.7, the heave and the development of heaving pressures will be restricted to the vicinity of the frost line. Consequently, the temperatures being only slightly below 0 °C, the suction and thus possible values of P_i will be relatively small.

Equation [7.10] shows why frost heave is less in 'well-drained' soils, where the initial pore water pressure P'_w is relatively low, or substantially negative. Conversely, large frost heaves are frequently observed in association with a high water table.

An alternative way of looking at these aspects is to consider the effective stress $\acute{o}$ existing *prior* to freezing. Effective stress is

overburden pressure minus pore water pressure, and a suction necessarily involves an equal effective stress (Chapter 4). If the suctions that would be developed by the relevant freezing temperature are less than $\acute{o}$ then there is no fall in pore water pressure, no migration of water, and no frost heave. The pressure of the ice P_i will not reach the overburden pressure, ice segregation will not occur, and the overburden will not be lifted.

In summary, pressures generated by frost heaving range from virtually nil in coarse soils up to several 10^5 N m^{-2} in fine grained soils, under seasonal freezing conditions. Wet conditions will give the larger pressures. Frost heave pressures only develop (as does any pressure) to the extent that there is an equal constraining or reactive pressure. The smaller the constraining pressure, the bigger the frost heave. Where freezing conditions persist perennially, higher pressures are to be expected, due to secondary heaving, but heave will occur relatively slowly. The maximum heaving pressures may not be generated therefore, against structures which can yield to some degree. While the process of heave and of the development of the heaving pressures is increasingly understood, it is frequently very difficult to predict maximum heaving pressures for engineering purposes.

7.10 Phase composition and practical procedures

The fundamental importance of transitions between ice and water, in the study of properties of frozen soils, is only gradually being realized. Increased geotechnical activity in cold regions has stimulated research. Testing procedures are being developed for the characterization of freezing soils with regard to their thermal and mechanical properties. Firstly, it is basic to know the proportions of ice and water present in the soil as a function of temperature. To this end the (unfrozen) water content can be determined by several methods. Calorimetry may be used, but is time consuming and fairly difficult. Dilatometric methods are perhaps the most accurate (Williams 1976) but are also difficult. For saturated materials the volume increase on freezing reveals the amount of water frozen. A relatively simple procedure is the use of suction-moisture content tests (carried out on unfrozen samples, the results from which are easily converted to unfrozen water contents (section 7.5). A promising new method involves a probe which measures the dielectric properties of the frozen soil (Patterson and Smith 1981).

Knowledge of the proportions of ice and water in a soil, as a function of temperature, permits calculation of thermal properties. For any temperature change (cf. soils in Fig. 7.5) the associated freezing or thawing necessarily involves a quantity of latent heat, which commonly exceeds the quantity of heat identified with the heat capacity of the soil components. The term 'heat capacity' for the freezing soil is normally replaced by 'apparent heat capacity' to recognize the latent heat component. Accurate knowledge of soil thermal properties is essential for calculation of depth of frost penetration, or of thaw of permafrost, and similar problems (see Chapter 11). Measurement or prediction of thermal conductivity in frozen soils is difficult. Conductivity is relevant to situations involving temperature gradients and these imply suction gradients and thus movement of moisture. This movement itself involves heat transfer. Johansen (1972, 1973, 1975 discussed in Chapter 5) has devised an estimation procedure for the 'thermal conductivity' of frozen soils as well as for unfrozen soils.

For most engineering purposes grain size composition is used as a guide to susceptibility to frost heave but clearly this cannot be completely satisfactory as the frost heave also depends on environmental conditions. Research in recent years has involved measurement of heaving pressures of samples which have access to free water at known pressures. Ultimately the question is one of coupled heat and moisture (energy and mass) flows, which, as indicated, involve complex theoretical approaches.

The mechanical properties of strength and deformation (see p. 179) are obviously related to phase composition and to changes in the proportions of ice and water. Such changes occur in response to stress, as well as to temperature. Although the changes in phase composition in response to 'externally' applied stresses (Williams 1976) may not be large, they are significant, and indeed appear to be a part of the deformation process. Slow migration of water leads ultimately to changes in distribution of ice, and is promoted by stress gradients as well as by temperature gradients.

7.11 Gibbs free energy of water in frozen soils

It was earlier noted (section 7.8) that the Gibbs free energy is an appropriate function to utilize in considering freezing of porous media, even though it is often more convenient, albeit with some loss of rigour, to consider suction, pressure or similar quantities relating to potential. Both because of its importance in more advanced studies and because of its widespread application to phase equilibria by physical chemists and others, we briefly consider the Gibbs free energy with respect to freezing soils. It may be noted that the term is here used to mean the free energy *per unit mass* (chemists commonly call this 'chemical potential' while they reserve 'free energy' for the non-specific sense (Penner 1968).)

At the freezing point of a substance, the free energy G_1 of one phase, equals that of the second, G_2. At temperatures below the freezing point, the free energy of the liquid is higher than that of the solid, and accordingly equilibrium does not exist. Instead the liquid is transferred, spontaneously, to the solid phase. On the other hand, liquid and solid can coexist at other temperatures than the 'normal' freezing point provided that the free energy of one or both phases is so changed as to again produce equal free energies in the two phases. A change in free energy dG of the water occurs of course, if the water content of a soil is decreased. Such a decrease happens in a frozen soil in association with water being transferred to ice. This explains the coexistence of quantities of water with ice in soils at temperatures below 0 °C.

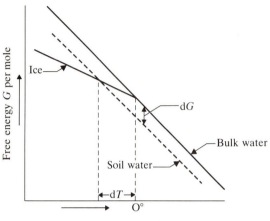

Fig. 7.8 Free energies of ice, bulk water and soil water as a function of temperature. Ice and water normally coexist at the freezing point 0 °C. If the free energy of the water is lowered, as in the case of the soil water above, the freezing point is changed by dT. Alternatively, the conditions may cause the free energy of the ice to rise, which will also result in a lower freezing point. (Modified after Everett 1961)

In Fig. 7.8 the free energies of pure 'bulk' ice and water are represented as a function of temperature. At the freezing point 0 °C they are of course equal. By changing the free energy of the water by some amount dG, at temperatures below freezing, the water is again brought into equilibrium with the bulk ice. The state of the water in frozen ground, its suction, constitutes just such a change. That state represents the effects of adsorption and possibly osmotic forces, and for temperatures near 0 °C, capillarity.

The value of the Gibbs free energy in a full understanding of many of the relationships discussed in this chapter and in Chapter 6, is amply demonstrated by the classic work of Edlefsen and Anderson (1943) – even though

curiously enough their detailed analysis of the situation of water and ice *in the soil pore* appears erroneous (Williams 1967 p. 21).

The symbol G has been used to refer to *absolute* free energy, a quantity we cannot actually measure. Just as with 'potential', when we speak of 'free energy' we are usually referring to a measured *difference* in free energy between our point of interest and some datum free energy, defined as that for example of a body of pure free water. If we wish to make this absolutely clear, we use the term *relative free energy*. In defining potential we used the word 'isothermally', or 'at the same temperature' in referring to the relationship with the datum. The datum for dG, is also to be taken as the free energy of pure bulk water G_{T_w} at the temperature T of the soil and at the same elevation. In the case of frozen soil (Fig. 7.8) the bulk water necessarily is hypothetical, in so far as the temperature is below 0 °C.

7.12 Temperature-induced moisture movements

Clearly, in many respects the free energy is equivalent to potential as considered in Chapter 6. But 'potential' as defined cannot be used to describe the effects of a temperature gradient (non-isothermal conditions) on water movement, and there is a related possible pitfall with respect to free energy. Reference to Fig. 7.8 may lead to the conclusion that, as there is a steady decrease in absolute Gibbs free energy with rising temperature – consider the slope of the line for temperature above 0 °C – there must also be a steady decrease in potential. One might suppose there to be a marked potential gradient associated with temperature gradients in unfrozen soils, such that there would be significant flow. But there is in fact, evidence that moderate temperature gradients in unfrozen *saturated* soil do not cause much movement of liquid water. The reason for the apparent paradox is that potential relates essentially to mechanical properties, forces and pressures. Obviously, warming the soil does not change the pressures or the equivalent suctions of the water in the soil much (there may be a small decrease in suction with increase in temperature). The Gibbs free energy implicitly involves temperature and the entropy property, as well as pressure and volume ([2.3]). When the Gibbs free energy is changed by changing the temperature this does not merely involve changes in the content of energy in the form available for work. Furthermore, our methods of assessing soil water potential or free energy involve a datum body of water at the *same* temperature as the sample. Obviously, comparison between soils at different temperatures is not possible.

In the case of unsaturated soils, where vapour movement is relevant, the situation is somewhat different, but equally illustrative. As noted in section 7.3, vapour pressure (and thus the free energy of the vapour) increases with temperature, and vapour therefore moves from *higher* to *lower* temperatures. But Gurr, Marshall and Hutton (1952) describe a situation where condensation of the vapour at lower temperature increased the moisture content (decreasing the suction) sufficiently to cause *liquid* flow in the opposite direction (i.e. towards the higher temperatures).

Difficulties can often be avoided by using the free energy function where appropriate, in studying phase transitions and other situations involving heat energy, and judiciously switching to potential, or even pressure, when necessary, in order to handle questions of a mechanical nature. So long as it is remembered that potential or free energy is measured relative to a datum body of water at the same temperature as the soil, confusion is normally avoided.

The relative complexity of some of the thermodynamic considerations must not lead to underestimating their importance for a proper understanding of many earth surface processes. But we cannot wait for a complete and comprehensive understanding of the thermodynamics of earth materials, before attempting analysis of natural situations. While many of the relationships touched upon in this chapter will be considered subsequently in relation to particular terrain features, others have been introduced as a link with studies in more basic science. Such linkages are the source of progress in our understanding of phase transitions and their significance in soils.

8

Microclimatological properties of ground surface

8.1 Surface conditions and the ground thermal regime

The temperatures and heat fluxes within a layer extending ten or twenty metres below the ground surface depend upon conditions external to the ground, in a manner which is modified by the nature of the surface itself. The earth's elevated internal temperature gives the geothermal flux of heat which provides a lower boundary condition for the layer, but the flux is so small that it can often be ignored. The fundamental element in the ground thermal regime is the sun's radiation, although only a small fraction of this energy enters the ground. The remainder is mostly reflected, involved in evaporation, or transferred to the air as outlined in Chapter 3.

Simultaneously the ground surface itself emits radiation. The source of this is energy stored within the ground. Often the energy storage is recent, as in night time cooling involving that energy supplied by solar radiation during the preceding day. The period of storage may be extremely short. Even as the sun warms the ground, the emission of radiation increases. The heat energy may also have entered the ground following condensation at the surface, by conductive transfer from the air, and other processes. Clearly, the magnitude of the flux of heat into or out of the ground depends very much on the nature of the surface. This flux and its variations, which follow from the temperatures established at the surface of the ground, are the essence of the ground thermal regime; such knowledge is very important in agriculture, increasingly so in building design and civil engineering, and so far as the associated transfers of moisture are concerned, in hydrology. Yet, mainly because the flux into or out of the ground, the flux G in [3.1], is normally small, compared with the other fluxes, it does not often figure prominently in discussion of the surface energy and mass exchange. Variations in the ground flux can only be understood in terms of changes in the other fluxes which are considered in this chapter. The subsequent disposition of the energy which enters the ground is considered in Chapter 10.

8.2 Radiation balance

The logical starting point for our considerations is the balance of incoming and outgoing radiation. By 'balance' we are here referring to a net incoming or outgoing quantity of radiative energy. This is equal to the sum of all other heat fluxes (whether positive or negative) according to the conservation principle, illustrated by [3.1].

Direct solar radiation is 'short-wave', and Fig. 8.1 shows the wavelength distribution of solar radiation. Solar radiation which is deflected (made 'diffuse') in the atmosphere,

provides daylight even though the sun may be obscured by clouds, or other shadowing objects. This is also short-wave. These two sources of incoming radiation to the surface, we call respectively Q and q.

Radiation of wave lengths 1.0 microns to 0.61 microns, and 0.51 to 0.31 microns is particularly important in photosynthesis and plant growth which consequently is affected by variations in spectral distribution. Such variations result from atmospheric composition and other factors. Photosynthesis utilises at most some 7 per cent of the total incident radiation.

Long wave radiation, I, is emitted by terrestrial substances, in an amount proportional to absolute temperature according to the Stefan-Boltzmann law:

$$I = \varepsilon \, \sigma \, T^4 \qquad\qquad [8.1]$$

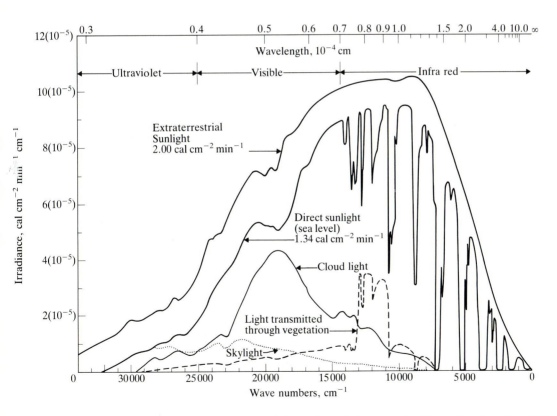

Fig. 8.1 Spectral distribution of extra-terrestrial solar radiation, solar radiation at sea level for a clear day, of sunlight from a complete overcast, and of sunlight penetrating a stand of vegetation. Each curve represents the energy incident on a horizontal surface. (From Gates 1965)

where T = the absolute temperature K of the surface

σ = a coefficient, the Stefan-Boltzmann constant

ε = the emissivity

The ground emits long-wave radiation, mostly in the range 4 – 100 microns, but also receives long-wave radiation from dust, moisture etc. in the atmosphere. The *emissivity* ε is the amount of radiation being emitted, expressed as a fraction of the maximum possible, that is, of the amount emitted by an ideal 'black-body' at that temperature. Moist soil surfaces have an emissivity between 0.90 and 0.95, most vegetation somewhat greater values, and snow 0.99. The emissivity varies somewhat with the wavelength of the emitted radiation. Much of the radiation emitted at the earth's surface is absorbed and reradiated back by water vapour, and carbon dioxide. This atmospheric counter-radiation reduces the night time cooling that would otherwise occur. We distinguish the incoming (atmospheric) long wave radiation as $I\!\downarrow$ and that outgoing from the surface as $I\!\uparrow$. The radiation balance of net radiation R is then expressed by:

$$R = \underbrace{Q + q - r}_{\text{Short wave}} - \underbrace{I\!\uparrow + I\!\downarrow}_{\text{Long wave}} \qquad [8.2]$$

The quantity r is that part of the solar and diffuse radiation which is reflected at the earth's surface.

The value of R may be positive (net income of energy by radiation) or negative (net loss of energy by radiation), the latter being the case during darkness. Obviously Q is affected by the sun's angle, and thus with latitude and time of day. $Q + q$ varies greatly with time of year and shading by topographic and other features is important.

Angle and orientation of slope is important with respect to direct solar radiation. Table 8.1 shows that at latitude 45° N incident direct solar radiation per day, in midsummer, is greatest on south-facing slopes (azimuth 180°) with an angle of about 15°. A north facing slope of the same angle receives some 12 per

cent less, while this reduction is some 22 per cent for slope angles of 30°. In midwinter the effects are much more marked, because of the lower solar altitude. The importance of latitude in this connection is thus evident. The effect of topography with respect to diffuse radiation is as might be expected, much less. The

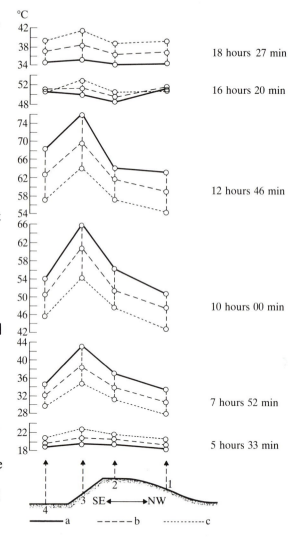

Fig. 8.2 Temperature of upper layer of sand in different parts of a barchan chain (a barchan is a ridge of sand found in deserts) (After Petrov 1976)
1. Windward slope
2. Crest
3. Slipslope
4. Interbarchan depression
a. sand surface; depth: b-1 cm, c-2 cm

importance of slope angle is well-illustrated by temperatures in a *barchan*, a sand ridge feature characteristic of deserts (Fig. 8.2). The absence of water or other microclimatic influences makes this example particularly illustrative.

The proportion of the incoming short-wave radiation which is reflected is given by the *albedo* or reflection coefficient of the surface. The value of r ([8.2]) is very variable, and changes for a given site with seasonal vegetation development, snow cover, and other surface conditions. The incident solar radiation is not solely in the visible spectrum, and the lightness of a surface is a poor guide to its albedo. Table 8.2 gives values for the reflection coefficient for all wavelengths in the range 0.3 to 4μ thus including ultraviolet and infra red. If a different range of wavelengths is considered somewhat different values apply.

Changes of several per cent occur in r, from hour to hour, mainly because the albedo, the percentage reflected, itself varies with the angle of the incident radiation, as well as with wavelength (spectral) composition (Fig. 8.3). Changes of moisture content significantly affect the reflectivity of bare soils (Idso *et al.*

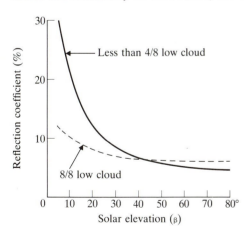

Fig. 8.3 Reflectivity of a plain water surface as a function of solar radiation and cloudiness. (After Monteith 1973)

1975) and wet sand, for example, has an albedo about one half that for the dry state (Geiger 1965). It is often assumed that the long-wave radiation $l\!\downarrow$ received by the surface is not reflected, that is, the surface behaves as

Table 8.1 Daily direct solar radiation on slopes of different angle and azimuth, latitude 45 °N, (after Wilson, 1975), in J cm^{-2}

8.1a Summer solstice
Atmospheric transmissivity: 0.60

Angle of slope (degrees)	Azimuth of the slope*																		
	0	10	20	30	40	50	60	70	80	90	100	110	120	130	140	150	160	170	180
0	2077	2077	2077	2077	2077	2077	2077	2077	2077	2077	2077	2077	2077	2077	2077	2077	2077	2077	2077
10	1964	1968	1974	1982	1992	2004	2017	2031	2045	2059	2073	2086	2098	2104	2108	2116	2123	2126	2128
20	1789	1792	1799	1811	1827	1847	1871	1897	1927	1956	1985	2013	2038	2060	2079	2094	2105	2112	2114
30	1561	1565	1575	1593	1616	1648	1687	1733	1780	1826	1869	1909	1943	1972	1996	2014	2027	2034	2037
40	1286	1290	1304	1326	1359	1412	1476	1544	1610	1670	1724	1771	1810	1840	1863	1881	1892	1898	1900
50	971.0	977.0	993.2	1020	1083	1171	1262	1349	1429	1499	1559	1607	1644	1672	1690	1702	1707	1708	1709
60	627.0	633.3	652.0	724.4	839.0	953.0	1060	1158	1244	1318	1376	1422	1453	1473	1481	1482	1478	1471	1469
70	279.0	306.0	394.0	519.0	645.0	765.4	876.1	975.2	1061	1130	1184	1221	1243	1250	1245	1231	1213	1197	1190
80	147.0	178.0	264.0	377.0	493.0	605.3	710.2	803.0	881.0	942.0	986.5	1013	1021	1014	992.0	961.0	925.0	895.0	883.0
90	88.20	114.0	182.0	274.0	372.0	469.4	561.0	641.0	707.3	757.0	790.0	803.4	798.0	775.0	736.0	685.1	628.0	581.0	563.0

* The azimuth of the slope in degrees from north through east.

8.1b Winter solstice
Atmospheric transmissivity: 0.80

Angle of slope (degrees)	Azimuth of the slope*																		
	0	10	20	30	40	50	60	70	80	90	100	110	120	130	140	150	160	170	180
0	457.3	457.3	457.3	457.3	457.3	457.3	457.3	457.3	457.3	457.3	457.3	457.3	457.3	457.3	457.3	457.3	457.3	457.3	457.3
10	220.0	224.0	234.5	252.0	275.5	304.3	336.9	373.3	411.3	451.0	490.7	529.6	566.0	599.0	622.4	650.4	667.5	678.0	681.3
20	13.0	20.0	41.0	76.0	120.4	172.2	231.2	295.9	365.8	438.9	513.3	587.3	658.0	722.3	778.7	824.3	857.7	878.2	885.3
30	0	0	0	0	32.6	88.2	157.6	238.7	328.5	425.9	527.9	630.8	730.7	824.3	906.2	927.7	1022	1052	1062
40	0	0	0	0	6.3	46.4	112.4	197.7	298.5	411.3	532.5	657.9	782.5	901.2	1006	1091	1155	1193	1206
50	0	0	0	0	1.3	27.2	84.4	167.2	271.7	393.3	510.8	668.4	812.6	950.6	1076	1177	1253	1299	1315
60	0	0	0	0	0.4	18.0	66.0	143.8	246.6	370.3	483.2	661.7	818.0	971.9	1112	1227	1313	1365	1382
70	0	0	0	0	0.4	12.5	52.2	122.9	220.7	342.3	444.8	637.9	800.9	964.3	1115	1240	1333	1389	1408
80	0	0	0	0	0	8.8	41.8	104.1	194.0	308.9	397.1	597.3	760.3	927.1	1084	1215	1312	1371	1392
90	0	0	0	0	0	6.3	33.0	86.1	165.5	270.4	17.10	541.3	699.3	862.8	1020	1153	1252	1312	1332

* The azimuth of the slope in degrees from north through east.

8.1c Equinoxes
Atmospheric transmissivity: 0.75

Angle of slope (degrees)	Azimuth of the slope*																		
	0	10	20	30	40	50	60	70	80	90	100	110	120	130	140	150	160	170	180
0	1545	1545	1545	1545	1545	1545	1545	1545	1545	1545	1545	1545	1545	1545	1545	1545	1545	1545	1545
10	1253	1257	1269	1289	1316	1349	1388	1431	1476	1522	1569	1614	1656	1694	1727	1754	1774	1786	1790
20	923.0	931.3	956.0	997.3	1055	1124	1203	1289	1378	1468	1557	1643	1723	1795	1858	1910	1949	1973	1980
30	565.1	578.1	623.2	697.6	792.9	901.6	1020	1144	1271	1396	1520	1638	1748	1848	1936	2009	2065	2099	2110
40	189.8	227.4	322.3	439.3	569.7	710.6	857.7	1009	1162	1313	1460	1601	1733	1853	1960	2049	2118	2162	2177
50	0	38.9	140.4	266.7	407.6	560.1	720.6	886.2	1053	1218	1379	1534	1678	1811	1929	2029	2108	2159	2177
60	0	10.0	68.55	170.1	299.3	446.4	606.5	722.5	943.0	1113	1278	1436	1585	1722	1844	1950	2035	2091	2111
70	0	4.2	38.9	115.4	224.5	357.4	507.5	666.7	831.4	995.7	1157	1312	1455	1590	1710	1814	1900	1959	1981
80	0	2.1	25.1	81.5	171.0	285.9	418.8	565.6	716.0	869.9	1020	1163	1297	1420	1531	1628	1710	1768	1791
90	0	1.3	16.3	58.1	128.7	224.9	340.3	466.9	601.9	736.1	870.3	995.7	995.6	1218	1313	1397	1469	1524	1546

* The azimuth of the slope in degrees from north through east.

Table 8.2a Albedo of natural surfaces*

Fresh snow	0.85 – 0.95
Old snow	0.42 – 0.70
Dry-bare sandy soils	0.25 – 0.45
Dry-bare clay soils	0.20 – 0.35
Bare peat soils	0.05 – 0.15
Most field crops	0.20 – 0.30
Forests, deciduous	0.15 – 0.20
Forests, coniferous	0.10 – 0.15
Forests, deciduous with snow on the ground	0.20
Tundra, wet, snowfree	0.17
Burned over	0.12
Water	0.03 – 0.10

(Modified from Rosenberg, 1974)
* A greater range than indicated may occur for certain categories of surface.

Table 8.2b Vegetation – maximum ground cover*

	Latitude of site	Daily mean
(a) Farm crops		
Grass	52	24
Sugar beet	52	26
Barley	52	23
Wheat	52	26
Beans	52	24
Maize	43	22
Tobacco	43	24
Cucumber	43	26
Tomato	43	23
Wheat	43	22
Pasture	32	25
Barley	32	26
Pineapple	22	15
Maize	7	18
Tobacco	7	19
Sorghum	7	20
Sugar cane	7	15
Cotton	7	21
Groundnuts	7	17
(b) Natural vegetation		
Heather	51	14
Bracken	51	24
Gorse	51	18
Maquis, evergreen scrub	32	21
Natural pasture	32	25
Derived savanna	7	15
Guinea savanna	9	19

From Monteith (1973) after various authors.
* Average values of albedo (%), vegetation-covered surfaces. Albedo is to some extent dependent on the angle of the incident radiation and hence on latitude of the site.

a 'black body'. In fact as much as ten per cent may be reflected from some surfaces.

Because the net radiation R depends on the nature of the surface and involves surface temperatures (controlling $I\uparrow$) and wetness (affecting albedo), it is itself a function of the surface energy and mass balances. Clearly the net radiation R can only be assessed with some approximation. Various instruments are available for the purpose (Rosenberg 1974). Pyranometers involve a black sensor the temperature of which is a measure of the solar and diffuse (shortwave) radiation, $Q + q$. A similar device with the sensor facing downwards measures the short-wave radiation reflected from the ground surface. The ratio of the observations is a measure of the albedo. Instruments known as net radiometers measure R directly, following similar principles but also accounting for the long-wave components $I\downarrow$ and $I\uparrow$.

A major complication arises when the net radiation R is required for the actual surface of the soil or rock. Most published values refer to a plane just above the vegetation layer. A further series of radiative exchanges takes place between the vegetation surface or surfaces, and the ground below. Figures quoted by Wilson (1975) show that R at a forest floor is often less than 10 per cent that at the top of the canopy. The figure varies with solar angle. The question is considered in more detail by Geiger (1965). The incident long-wave radiation $I\downarrow$ is also modified substantially, and includes a component arising from the foliage and branches although Geiger suggests this to be generally of little significance. Low vegetation such as grass also has comparable effects on the value of R at the ground surface depending on the density of the vegetation. It is only in the case of bare soil or rock, or perhaps of a relatively simple surface cover such as snow or artificial materials such as asphalt, that more or less direct measurements of net radiation at the ground surface can be made. These can be used as the basis of calculation of energy balance components, at the ground surface in the strict sense.

The incoming (downward) radiation depends on latitude, cloudiness, air pollution, humidity, daylight hours, and other factors of an atmospheric and astronomic nature. But when we consider the ground surface, we find that the innumerable local variations of vegetation and other surface features control the disposition of the incoming radiation. Ultimately, the geometry and composition of leaves and branches, and other objects covering the surface together with the directional nature of incident radiation determines the radiative exchange (Monteith 1973).

Leaf spectra

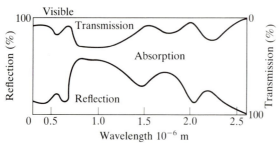

Fig. 8.4 Idealized relation between the reflectivity, transmissivity and absorptivity of a green leaf. (After Monteith 1973)

In Fig. 8.4 the relation between reflectivity, absorptivity and transmissivity, of a single leaf is illustrated, and is seen to be a function of the wavelength. A geometric effect, that of height of a crop, is illustrated in Fig. 8.5. In Table 8.2b average values of albedo are shown

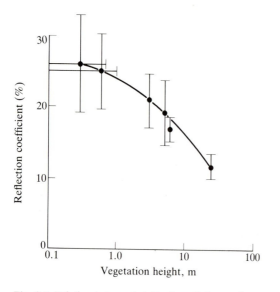

Fig. 8.5 Relation between height of vegetation and reflection coefficient. (After Monteith 1973)

for different crops. It is important to note the different scales of approach: an extremely *micro*climatic approach is necessary in investigating for example thermal stresses induced by solar radiation on porous rock. By contrast, studies of the energy balance of a field crop may utilize more generalized data (Table 8.2b). The information in Table 8.2b, also illustrates (as does Fig. 8.3) that while albedo is certainly to be regarded as a property of the surface, it is modified by external factors. The effect of latitude presumably relates primarily to solar angle.

8.3 Evaporative and sensible heat fluxes

The terms LE and H in the heat balance equation [3.1] refer to the heat associated with evaporation or condensation, and with warming or cooling of the air, respectively. The 'evaporative-condensative' flux involves the quantity of heat, 2454 J g^{-1} at 20 °C, taken in by water on its conversion to vapour, or liberated on its condensation. When evaporation occurs from a living plant the process is known as transpiration, so the flux is often called the evapo-transpirative flux.

We are concerned with the extent to which these processes occur, at that surface to which the balance equation applies. However the flux necessarily depends upon the movement of vapour to or from the surface. The component LE is then best evaluated as the vapour flux, E g cm^{-2} s^{-1}. If such movement did not occur, the vapour pressure in the vicinity of the surface would assume the equilibrium value corresponding to the condition (notably the temperature) of the water. Condensation or evaporation would cease. This equilibrium vapour pressure depends on the free energy (or potential, if isothermal conditions are considered) of the water as discussed in Chapter 7.

Falling temperatures as at night often result in the vapour pressure of the air becoming equal to the saturation vapour pressure (as the latter falls with temperature), and moisture is then condensed from the air as free water (dew). The surface (of the plant or soil) may have a different temperature to the air in general, and the relevant temperature determining whether evaporation or condensation occurs is that of the surface. The evaporative-condensative heat flux term has a positive sign for condensation (that is, heat is transferred to the surface).

The transport of both vapour and heat occurs primarily in association with movement of the air in the turbulent manner. The movement of vapour to or from the ground surface by molecular diffusion through static air is small. Because of the low thermal conductivity of air, transfer of heat by simple conduction alone through static air also would result in an extremely low flux. There is, however, a significant transfer of both vapour and heat by molecular diffusion in a very thin (a mm or so) layer of air adjacent to any solid surface where high temperature gradients or vapour concentration gradients give a relatively high diffusion flux through the thin layer. This more or less stationary layer persists even though air may be moving over the surface due to wind, and it is due to friction between the air and the surface, and to viscosity of the air.

Turbulent flow involves a series of rotary motions or eddies of the air. Laminar flow in which particles, such as in cigarette smoke in a still room, are seen to be moving in more or less straight lines occurs only at extremely low air speeds. Outdoors this is the case only in a few special situations, such as within the millimetre-thick boundary layer of solid surfaces mentioned above.

For an initial consideration of turbulent transfer we can ignore *advection*, the lateral transfer of heat of vapour, parallel to the plane of interest. If strictly parallel, such a flux has no component normal (i.e. perpendicular) to the surface, and thus has no *direct* effect on the perpendicular fluxes, with which we are now concerned. In other words we will initially

assume the perpendicular distribution of temperature, vapour pressure etc. as constant with time. This assumption does not imply, of course, that there is no wind, but merely that the wind is parallel to the surface. Turbulence due to wind is forced turbulence.

8.4 Forced turbulence

If we consider wind moving over the ground surface in a horizontal direction, there will be a drag on the lower layers of air exerted by the vegetation, and other irregularities, this constituting the roughness of the surface (considered further in section 8.7). The drag is not to be regarded as due to viscosity of the air but rather to the effect of pressure differences on lee and windward sides of obstructions, which tend to produce a force in a direction counter to that of the wind direction. The tendency to maintain a steady state results in some faster-moving air travelling downwards into the slower moving layers, while slower moving air is displaced upwards. This constitutes forced turbulent mixing and extends 500 m or so into the atmosphere. Turbulent mixing involves the transfer of momentum (mass times velocity) of the faster-moving air to the slower-moving air in the lower layers. This transfer is referred to as the momentum flux. It is given by:

$$M_f = \rho_a K_m \frac{du}{dZ} \qquad [8.3]$$

where M_f = mean flux of momentum
$\quad$ g cm^{-1}s^{-2}
$\quad \rho_a$ = density of air g cm^{-3}
$\quad K_m$ = eddy transfer coefficient, for momentum, cm^{2}s^{-1}
$\quad \frac{d\bar{u}}{dZ}$ = vertical gradient of wind velocity cm s^{-1}cm^{-1}

u refers to the average velocity over a period of time of the (horizontal) wind, which is lower as the surface is approached. Z refers to the vertical direction.

Equation [8.3] concerns a mixing process, and when there is a vertical gradient of temperature the process must equally involve the transfer of that heat contained in the moving masses. This is described by a similar equation:

$$H = -\rho_a c_p K_h \frac{\partial \overline{T}}{\partial Z} \qquad [8.4]$$

where H = heat flux, J cm^{-2}s^{-1}

c_p = specific heat capacity of moist air at constant pressure p, J g^{-1} °C

K_h = eddy transfer coefficient for heat, cm^2 s^{-1}

$\frac{\partial \overline{T}}{\partial Z}$ = average vertical temperature gradient in the air °C cm^{-1}

The average temperature gradient must be used because the gradient is in fact continually changing in association with the eddy formation.

The vapour flux (g cm^{-2} s^{-1}) is given by:

$$E = -\rho_a K_w \frac{\partial \overline{q}}{\partial Z} \qquad [8.5]$$

where K_w = eddy transfer coefficient water vapour (cm^2 s^{-1})

$\overline{q}$ = mean specific humidity (g vapour per g moist air)

Alternatively:

$$E = -K_w \frac{\partial \overline{\rho}_v}{\partial Z} \qquad [8.6]$$

where $\overline{\rho}_v$ = mean vapour density (or absolute humidity g cm^{-3})

A similar equation gives the flux of carbon dioxide – a quantity often significant in biological consideration. To the extent the various fluxes refer to the same eddies, it would be expected that the eddy coefficents would be equal:

$$K_m = K_h = K_w \qquad [8.7]$$

This equality, referred to as Reynold's analogy, has been the subject of much discussion, and it appears in fact, only valid under thermally neutral conditions, which are considered in the next section.

8.5 Lapse rates and convective turbulence

According to the gas law, the volume of a gas is inversely proportional to its pressure. Consequently, if air rises it expands, because atmospheric pressure decreases with height. It is also found that, generally, the temperature decreases with height during daytime. This is particularly so in sunshine when the ground surface becomes warmer and transmits heat to the adjacent air. The decrease (typically of about 0.6 °C per 100 m but varying quite substantially) is known as the lapse rate. During the night the reverse is often the case, the increasing temperature with height being known as a temperature inversion. If the temperature of the lower layers of air are warmer, because air is then less dense, the warmer air would be expected to rise due to buoyancy. This constitutes convection, and because the air movement is turbulent, we have convective turbulence (also known as free turbulence, or free convection, to distinguish it from forced turbulence considered previously). For convective turbulence to occur, however, the vertical temperature gradient, the lapse rate, must exceed a certain value known as the *adiabatic lapse rate*. The reason for this can be understood when we remember that the air generally *decreases* in density with height because of the decrease of pressure. The decrease in density associated with warming of the lower layers will only cause sufficient buoyancy for upward movement, if it exceeds the effect of the pressure gradient.

In thermodynamic terms, the adiabatic lapse rate is that vertical temperature gradient at which a parcel of air would not exchange any heat with its surroundings if moved in a vertical direction. That is, the parcel would itself change temperature such as to have the same temperature as the air surrounding it. In expanding, a gas cools. The adiabatic lapse rate represents exactly those temperatures the air would assume on account of its volume change on being moved up or down. The adiabatic lapse rate has a value of about 1 °C per 100 m, it being assumed that the decrease in density of the air with height is that for an atmosphere at rest.

During daytime as the ground surface is warmed, the temperature gradient frequently exceeds the adiabatic lapse rate and convection occurs. The air is then said to be unstable. 'Thermals' are large convective air masses which rise to considerable heights-and are sought by glider pilots. Conversely under inversion conditions, as at night, convection does not occur and conditions are stable.

Equations [8.4] and [8.5] do not appear to describe heat and vapour transfer under conditions of free convective turbulence. One reason is that the rising air masses may behave as very large 'bubbles' or packets of air, without any thorough degree of mixing occurring within them (Priestley 1959). The vertical temperature gradient of the air and the associated air movements are important and complex elements in the energy exchange at the earth's surface. Fuller consideration from the meteorological viewpoint is found in Hare (1966) and Barry and Chorley (1976).

8.6 Methods for determination of evaporative flux

The values of the eddy transfer coefficients, under natural conditions, vary through some four orders of magnitude. While the relationships briefly described in section 8.4 and 8.5 (more detailed discussion is given in Rose 1966, Munn 1966), give some insight into the physical processes involved, they are frequently difficult to apply directly. Such methods of determining the fluxes of sensible heat and vapour by eddy diffusion equations are known as aerodynamic. There are a number of alternative methods.

In the case of vapour flux, the evaporative loss of moisture (and thus the heat balance component LE) may be determined with an apparatus known as a lysimeter. In principle, this involves an arrangement whereby a body

of soil, usually a cylinder about 2 m in diameter and 2 m depth, together with its vegetation cover, is continuously weighed. Although the situation of the cylinder is not entirely natural, with proper precautions a reasonably characteristic measurement of evaporation can be made. The soil surface should be level with that of the surroundings, and a monitored source of water to the cylinder base corresponds to the presence of ground water.

Another method involves the determination of the water balance of the soil. This requires continuous monitoring of the soil water suction at sufficient depths, and also of the soil water contents (Royer and Vachaud 1974). The procedure enables determination of the flux of water upwards to the surface (where it is evaporated) and involves equations of the type given in Chapter 10.

An alternative approach to determining LE and H is the 'energy balance method'. This involves, firstly, determination of R, the net radiation, and G, the ground heat flux. The difference $R - G$, is $LE + H$, from the heat balance equation, [3.1]. Now it is possible to apportion the sum $LE + H$ to its two components by use of *Bowen's ratio*, $\dfrac{H}{LE}$ (Bowen, 1928). The procedure requires acceptance of [8.4] and [8.5], and of Reynold's analogy, but it does not require evaluation of the coefficients K_h and K_w. We note that dividing [8.4] by [8.5]:

$$\frac{H}{LE} = \frac{c_p K_h (\partial \overline{T}/\partial Z)}{\ell K_w (\partial \overline{q}/\partial Z)} \qquad [8.8]$$

ℓ refers to the latent heat of vaporization, J g^{-1}, and appears because in [8.5], the flux E is in g cm^{-2} s^{-1}, while in [8.4], the flux H is in J cm^{-2} s^{-1}. LE, J cm^{-2} s^{-1}, represents a flux of energy equal to the mass of water evaporating, E, g cm^{-2} s^{-1}, times the latent heat, ℓ, J g^{-1}. Thus dividing both sides of the combined equations by ℓ gives [8.8] with $\dfrac{H}{LE}$ then being a dimensionless ratio.

Assuming $K_h = K_w$ as discussed above, and

$\frac{c_p}{\ell}$ having a value of about $4.2 \ 10^{-4} \ °C^{-1}$, then:

$$\frac{H}{LE} \simeq 4.2 \ 10^{-4} \frac{\Delta T}{\Delta \bar{q}} \qquad [8.9]$$

Bowen's ratio, LE, is usually given the symbol β. Its usefulness lies in the fact that it may be calculated if merely ΔT and Δq are known for a certain height interval ΔZ. Substituting into the energy balance equation we obtain:

$$H = \frac{R - G}{1 + 1/\beta} \qquad [8.10]$$

and

$$LE = \frac{R - G}{1 + \beta} \qquad [8.11]$$

For conditions of abundant moisture and with grass or crops covering the ground, Bowen's ratio has values from 0.1 to 0.3. Thus there is from one to three times as much energy involved in evaporation as in heating the air. However, under drought conditions, or at least a 'dry' surface, the ratio becomes much greater, tending to produce a larger error in LE calculated from [8.11], but a relatively smaller error in H calculated from [8.10]. Clearly a number of uncertainties attend the use of Bowen's ratio, not least those arising from the assumption of $K_h = K_w$.

A technique which has had substantial application in large-scale land management and hydrological studies, is the determination of the potential evapo-transpiration. This is the evapo-transpiration that would occur from a surface, the nature of which is optimal for evaporation. This is normally taken to be a vegetation-covered surface of certain characteristics and with abundant water available. If *similar* such surfaces are considered at various places, the evapo-transpiration will vary but in a manner which is essentially a function only of atmospheric (meteorologic) conditions. Of course the actual evapo-transpiration will depend on the nature of the surface at any particular place, but the potential evaporation is a valuable index for comparing the role of climates in evapo-transpiration.

The potential evapo-transpiration may be measured experimentally, for example with a lysimeter containing a surface of the prescribed characteristics. It may also be predicted by the use of empirically-derived equations involving climatic parameters. Such an equation was given by Thornthwaite (1948):

$$PET = C \ t^a$$

where PET is the monthly potential evapo-transpiration

t is the mean monthly temperature

C and a are constants

Other equations have been proposed, to include solar radiation and atmospheric humidity conditions. The approach tends to involve, implicitly, the assumption that potential evapo-transpiration mainly depends on the net radiation, at least in the absence of advection (section 8.8). There is, for example, no explicit recognition of the heat that may be supplied to the evaporating surface from within the ground.

Potential evapo-transpiration is the subject of a substantial literature (see e.g. Thornthwaite and Hare 1965) which attests its importance and also the inherent difficulties in the concept.

As an illustration of the latter, one notes that the potential evapo-transpiration is frequently greater than the evapo-transpiration occurring over a free water surface. The main reason for this apparent anomaly is that such a surface is normally cooler, and therefore has a lower vapour pressure than the 'optimal' surface.

Potential evapo-transpiration is not considered further here, as our primary concern is with the actual evapo-transpiration. Note however, that the ratio of actual evapo-transpiration to the potential evapo-transpiration serves as a useful measure of the relative efficiency of different ground covers as evaporating surfaces. Some examples of the role of different surface features in this respect are considered in section 8.9.

8.7 Vertical profiles of wind speed and surface roughness

The role of horizontal wind in producing a vertical momentum flux was described in section 8.4. It is the vertical momentum flux which is primarily responsible for the fluxes of vapour and sensible heat. Vertical flux of momentum depends on the wind speed gradient and thus on the vertical profile of wind velocity [8.3]. It follows that the fluxes of vapour (the evapo-transpiration flux) and of heat (sensible heat flux in the air) are also related to the wind velocity profile. A knowledge of the wind velocity profile is therefore basic to the aerodynamic methods, which use the eddy diffusion equations ([8.3] – [8.5]) for determining these fluxes.

For our purposes it is important to note the role of the surface cover, vegetation or other projections from the surface in modifying the vertical profile of the wind. This effect is referred to as the roughness of the surface, and the so-called roughness parameter is important in calculation of the fluxes.

The velocity of the wind, under neutral conditions (a vertical gradient of temperature corresponding to the adiabatic lapse rate) increases essentially linearly with the logarithm of the height Z. Below a certain height, this gradient of wind speed will show a dependence on the nature of the surface cover (Fig. 8.6). In the case of a smooth surface such as ice, wind speed falls to zero essentially at the surface itself. In the case of low, fairly dense cover, wind speed will tend to zero at height Z_0 somewhere above the ground surface proper. This height Z_0 constitutes the roughness parameter, and it is the point of intersection of the straight line, relating $ln Z$ and wind speed U (Fig. 8.6), with the $ln Z$ axis. It varies with vegetation height (Fig. 8.7). In the case of tall crops or other high vegetation, the wind speed

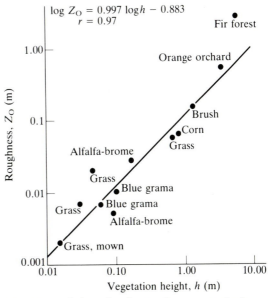

Fig. 8.7 Graph for estimating roughness parameter from vegetation height. (After Chang 1968)

versus logarithm of height shows a curvilinear form in that part within the height of the vegetation (Fig. 8.6). Thus the level of zero wind speed is in fact not at the height Z_0, but is displaced upwards by the amount D. This quantity is called the 'zero plane displacement'.

Quantitatively the role of the roughness and of the zero level displacement is quite difficult to elucidate. As would be expected the greater the roughness of a surface the greater is the mixing effect; Fig. 8.6 illustrates the steeper gradient, $\frac{dU}{dZ}$, associated with rougher surfaces.

In addition to this, the eddy transfer coefficients K themselves increase with wind velocity.

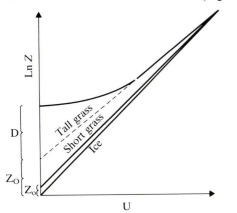

Fig. 8.6 Schematic diagram showing relationship between windspeed and logarithm of height. Z_0 is the roughness parameter; D is the zero plane displacement. (After Chang, 1968)

Where vegetation or other features give a layer of air (indicated by the roughness parameter) which is still, it is apparent that evaporation or heat exchanged at the ground surface proper will be reduced. Any analysis of the heat and vapour flux through a plane above the ground leaves uncertain the source of the fluxes, whether the soil surface, or plant members (which ultimately derive moisture from sub-surface layers). The existence of non-stable atmospheric conditions, and of advection further complicate the situation with respect to the wind speed profile, and the fluxes of vapour and heat.

8.8 Advection

Advection is defined as the transport of an atmospheric property solely by the motion of mass in the atmosphere; the term is generally used with reference to horizontal movement. Advection of particular importance is the lateral transfer of heat energy by wind. According to the conservation principle (as represented in [3.1]), if the net radiation arriving at the ground surface, does not equal the amount of energy leaving through evapo-transpiration and by conduction of heat into the ground, then the difference is supplied by the air (as 'sensible' heat).

If this state is continuous additional sensible heat must be being introduced, by the arrival of warmer air. Warmer, or colder, air often moves into a region for example warm or cool winds off the sea, or the movements of colder air into low-lying areas.

The effects may also be much more local. Sensible heat associated with warming of the air over certain crops or vegetation, may be transferred by wind to neighbouring areas of different surface characteristics. Sensible heat is often generated where conditions are locally drier, because of the smaller utilization of net radiation in evapo-transpiration. Instead, the temperature of the surface layers of the ground rise, and part of the heat gain is transferred to the adjacent air whose temperature consequently also rises.

Although advection phenomena are deliberately avoided in many field and theoretical studies of heat and mass balance, it is nevertheless a phenomenon of importance in a large number of situations (see for example, the discussions in Geiger 1965).

8.9 Nature of surface and water balance

Although the exchange of water between the ground and atmosphere is inevitably linked to the exchanges of heat, through the fluxes associated with evaporation and condensation, it is often convenient to prepare a balance equation for water arriving and departing from the surface region. This may be expressed as:

precipitation = change in soil moisture + evapo-transpiration + percolation + runoff

Precipitation includes all moisture of atmospheric origin: rain, snow, frost, dew and deposited fog droplets. If irrigation is practised, this also can be included for convenience as 'precipitation', and likewise there may be an input of water by surface flow from neighbouring areas. Part of this flow may continue out of the small area of consideration, then constituting part of runoff. 'Change in soil moisture' refers to water which is added or removed from the near-surface layers of the ground, changing their moisture content, while percolation refers to water proceeding further into the underlying ground. The 'percolation' term may have a negative sign if water is proceeding in the reverse direction. This apparently simple and illustrative equation requires considerable care in application. Another approach is to consider a plane at the ground surface, rather than a layer or layers of soil, with terms as follows:

Water arriving at surface from above ground sources P = evaporation E + water moving (in soil) to or from surface I + runoff on ground surface R

The term I can then be likened to the flux of heat energy in or out of the ground (G in [3.1]), and the consequent changes of moisture content and their distribution within the ground are determined by conditions in the soil as described in Chapters 6 and 10.

The incidence of rain falling on the ground surface is quite dependent on topographic relief. The effect of slope is, in general, to decrease the amount received per unit area, the effect also being dependent on the dominant wind directions and the non-vertical direction of fall of the raindrops. On a smaller scale, clumps of vegetation projecting above neighbouring cover will, under windy conditions, intercept more rainfall, often producing more favourable growth conditions. During light rainfall a substantial proportion may come to rest on plant surfaces, from which it subsequently evaporates. Humidity due to this water can reduce transpiration and passage of water from the soil through the plant. Of the rain which actually falls on to the ground surface little is evaporated directly, at least while rain is still falling, due to the high humidity of the air. If the soil is sufficiently permeable and not already saturated, the rain may infiltrate into the ground in its entirety. In heavy rainfall, less permeable soils will rapidly become saturated in the surface layers, and surface runoff will occur. Surface runoff is an agent of soil erosion, whether in the long-term geological sense, or in the agricultural sense of rapid loss of a significant part of the fertile layer (p. 144). A particular case of soils where the surface is largely covered by stones or boulders has been considered by Yair and Lavee (1976): In spite of the arid conditions in the Negev desert, water running off surface stones is collected together sufficiently to produce significant soil erosion.

Precipitation and surface runoff transfer heat to or from the ground surface, but are frequently overlooked in the heat balance equation. This illustrates again the linkage between heat and mass transfer, of which evapotranspiration is a further example. Evapotranspiration has received much attention from agronomists in relation to the water requirements of crops. Mechanisms involved in vapour transport have been considered in previous sections, and the following comments refer to the condition and availability of water for evaporation.

A basic initial approach is to assume that evapo-transpiration is proportional to the difference in relative humidities of the ambient air, and of the water at the evaporating surface. Water in soils and plants has a relative humidity dependent on the water potential as discussed in Chapter 7. Several qualifications are important. If as is often the case, the evaporating surface is at a different temperature to the air, then the difference in vapour pressure and not the relative humidities, is relevant (p. 93 and Fig. 7.2).

Because evaporation requires heat energy, the availability of such energy is a limiting factor. Indeed the decline or cessation of evaporation at night follows from the absence of incoming radiation, and the falling surface temperatures. Evaporation from the soil will cease some time in advance of the first formation of dew, when a soil surface is not saturated with water, because the vapour pressure of the soil is then less than saturation vapour pressure of adjacent air at the same temperature.

During periods of substantial evaporation a quite dry surface soil layer can develop, the relative humidity (or water potential) of which is often quite different from that measured even a few cm below the ground surface. Its hydraulic conductivity is low, however. The artificial provision of such a layer constitutes mulching, and is widely used to conserve moisture for plant use. The mulch protects the underlying surface from air and vapour movements, and thus retards evaporation.

The role of vegetation is complex. Evaporation from foliage is often considered in terms of a resistance to evaporation, governed by diffusion through tissues and the opening and closing of stomata which constitutes a physiological response to environmental conditions (Federer 1975). Compared to evaporation from free water surfaces there is certainly such a resistance, but in relation to the evaporation from the soil surface, the effect of plants is often to increase the total evapo-transpirative flux. Plant roots draw moisture from some depth below the surface, and continue to do so until a state of suction of about $15 \ 10^5$ N m^{-2} is reached, when permanent wilting occurs. The surface of the ground will often be dry by this time.

The nature of vegetation is also such as to produce large areas of evaporating surface. On the other hand, vegetation reduces the evaporative ability of the atmosphere through its effect in reducing wind speeds in the vicinity of the evaporating surfaces as noted in section 8.7. On the whole, vegetation generally both increases the evapo-transpirative flux and maintains higher soil moisture levels in the near surface layers than would be the case for bare soil. The surface layers are protected, while moisture for transpiration is drawn from greater depth by roots.

An approach developed by Baier and his associates in connection with growth of wheat has demonstrated the relation of 'available soil moisture' to evapo-transpiration. The available soil moisture is that part of the water content utilisable by growing plants. As it decreases so does the ratio of actual to potential evapotranspiration. The evidence (Baier 1969) is statistical, and presented in the form of graphs of percentage available moisture against $\frac{\text{actual evapotranspiration,}}{\text{potential transpiration}}$ also as a percentage. The relation depends on soil type. It is of value in a budgeting procedure relating climatic parameters and crop production.

8.10 Mass and energy balance components for specific situations

The complexity of the exchange processes, their continual change with time, and the difficulties of measurement mean that it is seldom possible to approach a complete description even for a single location or moment. In the case of relatively simple surfaces, such as that of a lake, intense study and experimental measurement has allowed the drawing up of a more or less complete set of values for the basic heat or mass balance equations, with the dependence of the component fluxes on environmental conditions being relatively well-defined. In other cases, values for the fluxes have been derived which, while more or less correct in a gross sense, do not refer to a specific point in the system represented by the ground surface. For example the heat and mass exchange may be determined immediately above a forest and in the ground immediately below it, perhaps with a fairly high degree of accuracy. Such determinations do not reveal the situation within the branches or around the trunks of the trees. Whether or not such 'simplified' determinations are of value depends on the purpose for which they are intended. If we wish to *modify* the atmosphere-ground exchange, a more detailed understanding may be necessary. An extremely 'micro-climatic' approach may be taken in studies of plant-water relations in which measurements, or theoretical determinations, are made with reference, for example, to a single leaf.

So far as we are concerned with the thermal and moisture regime of earth materials, it is clear that the role of complex surface cover, such as vegetation, is more problematic than that of heat and mass transfer within the ground itself. Simplification may extend to correlating observations of particular ground temperatures or thermal regime, in a purely qualitative manner, to the type of vegetation and other surface characteristics. Such an approach can easily lead to false conclusions, for example that the relative low temperatures under a particular cover are ascribable to high albedo of the cover, when in fact they may be due to a greater evaporative flux.

An energy balance study by Weller and Holmgren (1974) is an example of one in which the importance of particular surface characteristics have been established with some certainty, even though the terms of the energy balance equation have not all been analysed with great precision.

The study illustrates the importance of seasonal changes of surface conditions, at a natural arctic tundra site at Barrow, Alaska. Figure 8.8 shows the energy balance and the magnitude of the component fluxes on six

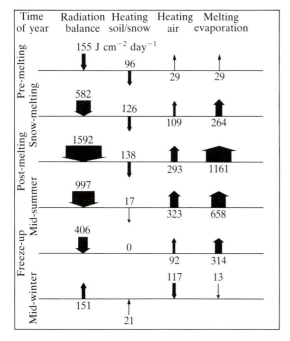

Fig. 8.8 Heat balances for six different times at a site in Alaska. The arrows show the direction of the energy flow and whether the energy path is in the ground, or above the surface (shown as horizontal line). The figures show the magnitude of the fluxes. (From Williams 1979 after Weller and Holmgren 1974)

occasions during the year. The most dramatic changes in the balance occur during the spring when the high albedo of the snow is replaced by the relative low albedo of the soil or vegetation. As a result the radiation absorbed increases greatly, although only 9 per cent of the energy is used to heat the soil while 73 per cent is involved in evaporation, following complete disappearance of the snow cover. The components of the heat balance on a daily basis during the snow-melt period are shown in Fig. 8.9 where heat arriving at the surface is

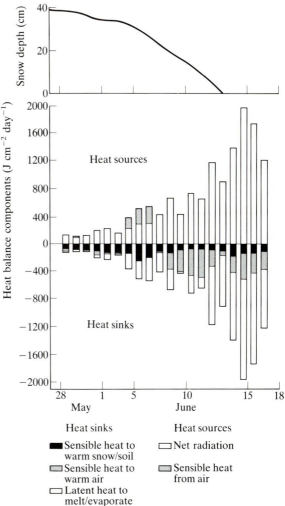

Fig. 8.9 Daily heat balances through the snow melt period. The top diagram indicates the snow depth during this period. (After Weller and Holmgren 1974)

given a positive sign, that leaving, a negative sign. The decrease of albedo occurs gradually, as the snow cover thins and there is an accumulation of dust and changes of crystal form at the snow surface.

Even before the snow has disappeared there is a heating of the air, which is perhaps explained by the transfer of heat from already bare patches being warmed by the sun. The effect of surface cover, snow and vegetation on the temperature of the soil surface at different times of the year, is show clearly in Fig. 8.8. The temperature gradients (Fig. 8.10) show that

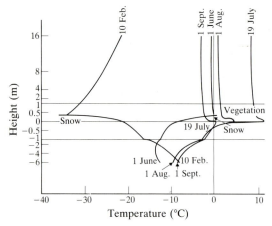

Fig. 8.10 Typical temperature profiles in air, snow, vegetation and soil. Note the effects of snow (10 February and 1 June) and vegetation (1 August and 19 July) on the soil surface temperature. The vertical height scale from +1 to −1 metre is expanded. (After Weller and Holmgren 1974)

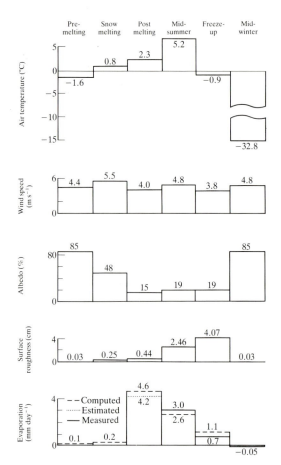

Fig. 8.11 Evaporation rates and factors that affect it, for six periods of the year (After Weller and Holmgren 1974)

on most occasions the soil surface is warmer than the vegetation or snow surface. At all times of the year the heat flux into, or out of, the ground represents nevertheless a small part of the net radiation. Note that in Fig. 8.8, the flux into or out of the snow is included together with that in the soil.

Several other quantities relevant to the surface conditions are shown in Fig. 8.11. Of particular interest are the evaporation amounts. Evaporation is low during the snow melt period. One factor in this perhaps unexpected observation, is that the air temperature is, during this period, substantially greater than that of the snow surface. Consequently the vapour pressure of the air often rises above the saturation vapour pressure for melting snow. At such times condensation on the snow takes place. The liberated latent heat causes further melting.

The results reported by Weller and Holmgren, and the fact, as they point out, that at other localities with superficially similar conditions, quite different quantities have been observed, illustrates the difficulty in generalizing about microclimatic effects. Direct observation, over extended periods of time, is normally necessary to establish in detail, the magnitudes of components of the energy or

water balance of a particular microclimatic environment.

For practical purposes it may be necessary to know the effects of deliberate or accidental surface changes, on for example soil temperature. The effects are not easily predicted. Shelterbelts, and the effects of living and non-living windbreaks, for example, have received considerable attention, but their effect on soil temperature is variable and often uncertain (Rosenberg 1965). The reduction of wind speed leads to reduction of evaporation, thus it might be thought, leaving more heat for warming the soil. In fact, in the sheltered areas, the transpiration behaviour of the plants is modified, the moisture regime of the soil is changed, and indeed the whole energy balance is modified in a complex fashion. A simple change such as decreasing the albedo may not lead to the increased absorption of solar energy expected: as the ground surface warms, the outgoing long wave radiation increases according to [8.1].

The need for detailed scientific analysis is evident. It is emphasized by the findings of empirical studies showing the importance of the nature of the ground surface in a more general sense. Pollution, the urban environment, agricultural and forest environments, and human health and well-being in general are all topics in which the microclimatological properties of the ground surface assume great importance. Good reviews are given for example in Mather (1974), Barry and Chorley (1976), Sellers (1965) and Ryden (1981).

PART III

Analysis of field situations

9

Slopes

9.1 The nature of slopes

Even to the casual observer topographic relief is related to the nature of the subsurface materials. The association of mountainous regions with long and steep rock slopes, and the association of plains with deep loose sedimentary soil material, are valid generalizations. On the local scale, however, the landscape is more confusing. Steep if not long slopes also occur, along rivers, in areas classed as plains. In high mountain areas, there are local occurrences of sedimentary material in terraces, plains or plateaux, which may be delimited by steep slopes composed of the same material.

To 'explain' these circumstances as different stages in denudation, as was attempted by early geomorphologists, or as the results of particular landforming agencies such as glaciers, wind or flowing water, has proven of singularly limited value in geotechnical and other applied studies. In considering the stability of slopes for engineering purposes, the engineer has turned his attention directly, if sometimes too exclusively, to the properties of the component materials.

How is the role of the properties of the earth materials to be related to that of external agencies such as glacier ice, or rivers, in controlling the form and stability of the slopes? The glacier ice may, as in the case of some depositional features, for example eskers, merely have provided a lateral support which was removed on deglaciation. Deformation or movement followed, giving the slope form we see today. The action of rivers and streams in relation to the formation of valley sides is also usually indirect, involving removal of support in a different way, by erosion at the foot of valley sides or river bank slopes where the water is in contact. The subsequent flows, land slides or other slope forming mass movements, follow from this erosion. But the nature of the movements and the form they give to the slope are not a unique effect of the river. The particular form of the majority of slopes depends upon properties of the ground materials themselves, and the manner in which they move, break or deform in the absence of support.

Some slopes of course, do owe their form largely to the direct action of some external agent, examples being rock slopes eroded and smoothed by ice, the water-smoothed barely sloping surface of river terraces, or the almost vertical sides of waterworn canyons – which, however, also enlarge subsequently by successive falls of rock. Even in these cases the direct erosive effects of glacier ice and river waters are themselves to some degree modified by the nature of the earth materials.

Some slopes owe their form to the erosive action of periodic flows of water over much of the surface during particularly heavy rains. Such surface run-off occurs because the soil conditions are such as to prevent immediate infiltration. In other respects too, these slopes represent a special group, where a somewhat different set of soil properties is important in mass movements.

9.2 Relative stability of slopes

A slope formed by an external agency, a steep glacier-smoothed rock face for example, may well be extremely stable. So also may be a slope formed by transient erosion of the ground surface by run-off. By contrast, a slope not so formed, is probably one which is the product of instability in the slope itself. That it stands at all is due either to the stabilizing effect of the latest movements, or to a change of conditions within the slope since the latest movements took place.

In Chapters 2 and 4, movement of materials on slopes was shown to result from the shear stresses exceeding the resistance, or shear strength, of the material composing the slope. Movement occurs when the ratio of shear strength to shear stress falls to below 1. The existence of a ratio greater than 1 implies on the other hand a stationary condition, the stability increasing with this ratio. In

geotechnical engineering the ratio is known as a factor of safety, F.

In the case of the ideal, infinite straight slope on dry non-cohesive material, the factor of safety has the same value anywhere in the slope. In nature of course, such a slope does not exist, although in certain situations conditions approximate it. In most slopes the ratio of shear strength to stress varies from point to point. The factor of safety *for the slope* refers to the sum of shear strengths and to the sum of shear stresses effective over the potential surface of movement or sliding surface. The sliding surface is often curved, many landslides involving the rotation of a mass of soil sliding over a more or less uniform arc (as seen in profile). The sliding surface may also be quite irregular depending on soil conditions. The determination of a safety factor of a slope before movement occurs is quite difficult, particularly because it is necessary to find the eventual sliding surface, that is the surface giving the lowest ratio of shear strengths to shear stresses. That ratio is then the factor of safety for the slope. In practice, computers are used to calculate a large number of 'trial' sliding surfaces, and a process of iteration is used to find the most likely one.

Quite frequently, a well-defined sliding surface is not developed. Instead, there is a plastic or viscous type of deformation. The stability analysis nevertheless may involve the assumption that sliding surfaces exist.

The conditions in a slope are not uniform with time. The effects of rain for example change conditions significantly over a short period. Consequently there is no single factor of safety in a precise sense. When engineers speak of a safety factor they often refer to a value calculated on the basis of the worst conditions thought likely to occur in some reasonable time, for example the life of a structure. In this sense the term often refers to a more uncertain, and more subjectively determined quantity. The distinction draws attention to the role of changing environmental conditions, and changing conditions within slopes in relation to stability.

Another qualification arises because of the different kinds of mass movement that can occur even in a single slope. A steep rock face for example may be completely stable (have a high safety factor) with respect to large rock falls or slides. Yet the same slope may be the site of constant small rockfalls, the anaiysis or prediction of which would require a quite different approach. Many soil slopes are stable with respect to rapid displacements, but are subject to slow, more or less continuous, creep movements. The latter are often difficult to analyze, and the concept of a safety factor with respect to creep is a particularly difficult one.

9.3 Role of friction and porewater pressure in slope form

An understanding of the role of friction and cohesion represented in the strength equation:

$$S = \sigma \tan \phi + C \qquad [9.1]$$

provides a basis for understanding slope form and stability.

The simplest example is again that of the dry, cohesionless soil, where $S = \sigma \tan \phi$ Consider a straight slope of such material. The shear stress (parallel to the slope) is (p. 47):

$$\tau = \gamma Z \sin i \cos i \qquad [9.2]$$

where Z is the vertical depth, i is slope angle, while the normal stress is: $\sigma = Z \gamma \cos^2 i.$ $[9.3]$ Dividing [9.2] by [9.3], and rearranging, gives:

$$\tau = \sigma \frac{\sin i}{\cos i} = \sigma \tan i \qquad [9.4]$$

For the slope angle i_c at which movement just occurs, shear stress τ = shear strength of soil, S, and we have:

$$\tau = S = \sigma \tan i_c$$

Comparing this equation to that for strength $S = \sigma \tan \phi$ [4.2], we see that:

$$i_c = \phi$$

Thus, the steepest angle at which dry, non-cohesive material will stand is equal to the angle of internal friction ϕ of the material. The angle of internal friction of dry, granulated sugar can be measured roughly by pouring it carefully from a bag into a pile. The uniform angle of a scree slope is similarly explained.

The kinetic energy possessed by larger stones falling on to the scree may actually result in a somewhat lower angle, since this energy is also involved in the adjusting movements of the material. On some scree or talus slopes therefore, there is an associated sorting of boulders by size, larger boulders lying lower down, in a part of the slope of somewhat smaller slope angle.

Other points to note for *dry, non-cohesive* material, are that the slope may have any length and remain stable. Reference to the equations also indicates that the depth, Z, of the plane of sliding is immaterial. The ratio of strength to stress is similar for all depths, as noted in the previous section. Thus the total depth of soil does not affect the maximum angle the slope may have. In fact, in nature (where conditions are never entirely uniform) the movements are often shallow, presumably because there is then least 'compressive' resistance at the lower end and least tensile strength resistance at the upper end of the moving mass (which will normally be only a part of the slope).

The situation is different for wet material. Consider again a long straight slope, but with a water table parallel to the surface. Water may be held above the water table by capillarity, of course, and there will also be movement of water in the soil in a downslope direction.

Now consider the strength of the soil, height H_1, *above* the water table (Figs 4.2, 4.3). It is:

$$S = [\gamma Z - (-\gamma_w H_1)] \cos^2 i \tan \phi \qquad [9.5]$$

The water pressure is negative (it is a suction), and the strength is therefore higher than for the dry material (p. 51). Consequently slopes composed solely of material having negative pore pressures will be stable at *higher* angles. This is why sand castles can be built – with damp sand.

At the level of the water table, the water pressure is 0, and $H = 0$. So far as the possibility of shear along the plane of the water table, we might conclude, therefore, that this would occur when $i = \phi$, as in the dry case. In fact, it is improbable that shear will occur along that plane, because failure of the slope is likely at values of i less than ϕ. At a distance H_2 below the water table, the strength in the wet soil at a depth Z is:

$$S = (\gamma Z - \gamma_w H_2) \cos^2 i \tan \phi \qquad [9.6]$$

Noting the effect of H_2 (compared to $H = 0$) it is apparent that soil at such a depth will shear with a slope angle i less than ϕ. It follows that the maximum angle the slope can have decreases as the thickness (total depth) of the soil increases. Consequently when slides do occur they will tend to involve a substantial thickness (depth) of material. If follows too that there will be a maximum height at which a slope is stable (compare Fig. 9.1).

Consideration of the equations also shows that the maximum stable slope angle is less, the higher the water table in the soil. This of course, agrees with the common observation that landslides tend to occur in wet periods.

The occurrence of the water table at the surface of the ground represents a special situation, in that the ratio of shear stress to strength is again constant with depth (assuming uniform conditions as described). With this situation:

$$S = (\gamma Z - \gamma_w Z) \cos^2 i \tan \phi \qquad [9.7]$$

the soil is either stable or, in the event, becomes unstable, simultaneously at all depths (assuming the material is uniform).

The maximum angle i_c the slope may have is then given by:

$$\tau = \gamma Z \sin i_c \cos i_c = S = (\gamma Z - \gamma_w Z) \cos^2 i_c \tan \phi \qquad [9.8]$$

This simplifies to:

$$\sin i_c = 1 - \frac{\gamma_w}{\gamma} \cos i_c \tan \phi \qquad [9.9]$$

and

$$\tan i_c \simeq \tfrac{1}{2} \tan \phi \ (\text{because } \frac{\gamma_w}{\gamma} \simeq \tfrac{1}{2}) \qquad [9.10]$$

This angle i_c often represents an ultimate value for the slope, on the assumption that at least occasionally the water table rises to the surface. Observations are reported of a number of slopes, to be described later, the angles of which have a tangent approximately equal to $\frac{\tan \phi}{2}$, and for these slopes it must be assumed that climatic conditions have, at sometime, been such as to have produced a water table at the ground surface. Carson and Petley (1970) apply this analysis to explain observed slope angles for various cohesionless materials, also pointing out that ϕ varies with the fineness and thus the degree of weathering. It follows that the value of i_c will change with time.

The foregoing discussion illustrates basic principles and their application under idealized conditions, for dry, and wet, slopes of non-cohesive materials. In nature, of course, soil material commonly varies with depth, there may be transient high water pressures in part of the profile (for example, in soil lying between impermeable layers) and many other complications. Some of these are considered in later sections.

9.4 Role of cohesion in slope form

The existence of a cohesive component of the soil strength has a fundamental importance, about which certain general statements also may be made. If, with some simplification, we regard cohesion as *constant* with depth, then the ratio of the shear strength, $S = C + \acute{o} \tan \phi$, to the shear stress, $\tau = \gamma Z \sin i \cos i$, always decreases with depth. Consequently, the possibility of sliding increases steadily with depth. Alternatively it increases with the height of a slope (Fig. 9.1). Slides, when they occur, will tend to be deep-seated. The maximum angle at which a slope of cohesive material is stable, is less as the soil layer is thicker, or as the slope is higher. A similar statement was made of course, about the slope in non-cohesive material and where there is a water table *below* the surface. For two reasons, however, the effect is more marked in the case of the cohesive soil. Firstly, cohesion is often a major part of the strength of such soils, particularly of course when the effective normal stress is low. It is well known in civil engineering, and indeed almost a matter of everyday experience, that vertical-sided excavations are possible in clay, to a shallow depth, because of cohesion. But if the excavation proceeds to greater depths the sides will inevitably collapse. This is because the shear stresses increase (Fig. 9.1), but the strength much less so. Secondly, for slopes of cohesive soil there is not an easily-defined limiting angle at which the slope is necessarily stable however great the depth of soil or height of slope. Only if the cohesion ceases to be effective (that is, is lost, as is quite often the case, see section 9.7), is there such a limiting angle. It is that given by [9.10]:

$$\tan i_c \simeq \frac{\tan \phi}{2}$$

These rather general considerations help us to understand the forms taken by landslides, and thus of the slopes themselves, in different materials.

It is also important to note certain qualifications. Both cohesion and friction angle ϕ are modified by the effects of water content and pore-water pressure. Thus in any detailed analysis, it is frequently not possible to regard $\tan \phi$ and C as constants having a single value for the material in question. Rather it is common to see the terms $\tan \phi'$ and C' where the suffix indicates that the values are appropriate to the relevant effective stresses and the associated water contents, and degree of consolidation. There is indeed much esoteric discussion in the literature both as to the true nature, and the manner of variation, of the strength components. Such considerations are often very important where precise quantitative data relevant to specific soil behaviour is

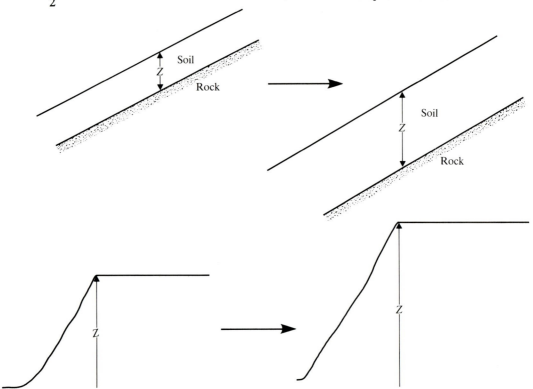

Fig. 9.1 In cohesive soils the shear strength does not increase with depth at the same rate as the shear stresses. Thus either a greater depth of soil cover (above), or a greater slope height (below) with a corresponding greater depth, results in increased chance of landsliding. This is also the case where a water table lies below the surface

required. But they need not deflect us from some more general conclusions concerning changes of cohesion in association with weathering and certain other natural processes. First, however, we consider the many geometric forms associated with landslides and other mass movements.

9.5 Geometric forms of displaced material

Most natural slopes differ greatly from the 'ideal' slope considered earlier. Not only is the surface profile often quite irregular, the materials within the slope vary. The geological history, of course, aids interpretation of the different layers often present, and of their spatial arrangement and orientation relative to the surface.

An excellent review is given by Skempton and Hutchinson (1969) of the range of forms exhibited by material displaced in mass movements, and their relationship to particular subsurface materials and structures. Although their review is directed primarily towards clay slopes, certain of the forms illustrated in Fig. 9.2, also apply to movements involving rock, or non-cohesive soils. There is no strictly standardised terminology, and terms such as 'slab' are used with somewhat different meaning by other authors.

The requirements for a 'fall' (Fig. 9.2) are an initially very steep slope, for example a rapidly eroded, (or perhaps excavated) clay face, or a rock face. A tension crack develops and movement, or failure, may occur by a surface of shear developing at the base of the column. This surface makes an angle to the external surface. On the other hand, if the tension crack (or perhaps, vertical joint plane in rock) extends to the bottom (Fig. 9.2), failure is essentially the overcoming of the compressive strength of the material under the weight of the column. Water pressure in the crack or joint clearly increases the chance of collapse.

Curved, often more or less circular, sliding

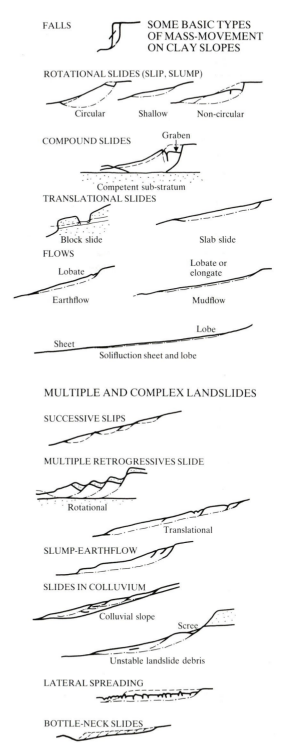

Fig. 9.2 Morphology of various forms of landslides and other mass movements. (After Skempton and Hutchinson 1969)

surfaces are common in clays or shales, and involve a rotation of the sliding mass. The curved form may be interrupted in 'compound' slides, where the sliding surface in part is the upper surface of a layer of stronger material, perhaps bedrock.

Normally the cohesive strength is responsible for the curved form. As noted, in slopes in cohesive material the height, or depth, is a limiting factor. The slope usually has a form (Fig. 9.3) such that the shear stresses are distributed in a manner resulting in the safety factor first being exceeded along a curved surface. Such rotational slips are analysed by determining moments of rotation; the ratios of resisting to rotating moments may be simplified in appropriate cases to give the equation for the safety factor as in Fig. 9.3. In very strong material more or less vertical slopes are common (the vertical sides of rock canyons, or the steep sides of glaciated valleys are examples). These also have a limiting height, but when they fail more or less plane sliding surfaces are developed. In this case the explanation is that the geometry, and the resulting stress distribution favours the safety factor being exceeded along a plane rather than a curved surface. Carson and Kirkby (1972) review this point in more detail.

Movements occur along shallow planes more or less parallel to the surface in slopes which

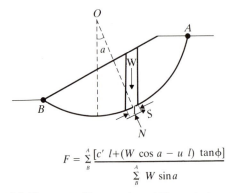

$$F = \sum_{B}^{A} \frac{[c' \; l + (W \; \cos a - u \; l) \; \tan\phi]}{\sum_{B}^{A} W \; \sin a}$$

Fig. 9.3 Diagram to illustrate the stability analysis of a deep-seated slip: the conventional method of slices. Notation: W = weight of slice; N = total normal force at base of slice; O = centre of circle which includes AB; S = shear strength at base of slice. (After Carson and Kirkby 1972)

approximate the 'ideal' slope, in non-cohesive material. The essence of this form of displacement, is that the depth of the sliding surface is small in relation to the length of the slope. That is, if the affected part of the slope is considered as being composed of vertical columns (Fig. 2.2) for the purpose of analysis, they will all be essentially similar. Note that in the 'method of slices' used by engineers to analyse deep-seated slides (Fig. 9.3) each column, or slice, is clearly different.

The scree slopes discussed in section 9.3 are an obvious example where 'ideal uniform non-cohesive' slopes are approximated, but most shallow, translational slides, with the sliding surface more or less parallel to the surface, in fact arise because of a sub-surface heterogeneity. The sliding surface may for example, be the underlying bedrock surface. The difference in material can also have a more indirect effect than that of this simple case of a layer of underlying inherently stronger material. Quite frequently weathering, which of course affects a layer more or less parallel with the slope surface, or higher pore water pressures in a layer with impeded drainage provide the explanation of the location of the sliding surface.

Once the landslide or other movement has commenced, it will continue until the shear stresses are again insufficient to overcome the strength of the material. We note in passing that as a result of movement, the strength of a material is often reduced, such that the geometry of a slope after stability is restored is likely to be substantially different. In some cases movements may involve such a velocity that kinetic energy moves the material much further than if the movements are relatively slow. Large rock slides for example, have often spread some way up the opposite side of a valley.

Quite frequently, material in the original slope is left in almost as unstable a situation as before, for example, by the formation of a steep slope in the upper part of a landslide 'scar'. As a result further landslides occur either almost immediately, or after a certain interval. The following landslide may, for example, be delayed until seasonal rainfall or other circumstances occur to produce the critical instability. Successive, discrete landslides or slips may occur (Fig. 9.2), or alternatively, a multiple retrogressive slide in which the originally moved material is further displaced along with additional masses (Fig. 9.2).

Such slides are common, for example, where a sudden and extreme loss of strength occurs in a particular layer. Such sudden loss of strength is characteristic of so-called quick clays, and much of the material may flow, the material continuing to move even though the slope is very slight. Certain essentially non-cohesive materials, sands, silts, and loess, may also become fluid in a process known as spontaneous liquefaction. While the circumstances leading up to liquefaction in the latter materials are different, most flows in the more or less liquid state, are ascribable to a collapse of the soil particle arrangement. The mineral particles instead of supporting each other, move in such a manner that their weight comes to be carried by the pore water. The resulting rise in pore water pressure, and reduction of effective stress, causes the abrupt loss of strength.

To complete this review of the kinds of movement, we refer to a variety of rather shallow movements which give rise to diverse and sometimes inconspicuous surface features. These include solifluction phenomena, which involve the effects of freezing and thawing. Their relative shallowness merely reflects the limited depths affected by the freezing and thawing process, but the particular, and sometimes as yet inexplicable cause of the movements suggests they be considered as a special group (see Chapter 11). A further quite distinct type concerns purely surface erosion effects due to ephemeral running water. The study of these involves principles applied in fluvial geomorphology, and they will be considered briefly, in section 9.10.

Finally there are two more or less distinct groups of slow and often somewhat continuous creep movements. 'Creep' may refer to small downslope displacement of material in the near-surface layers where climatic effects involve volume changes (shrinking and swelling, thermal contraction and expansion etc.). The word is also applied to very slow movements or deformations at any depth, occurring at stresses well below those necessary to initiate a sliding surface or other abrupt failure.

9.6 Changes of stability with time

If we ignore, for the moment, such creep movements, and consider earthflows, landslides, and other movements in which there is a sudden, and rapid displacement of a well-defined mass, it is apparent that we are often concerned with critical or 'triggering' events. For example, a slope which has stood through changing weather conditions, for centuries or more, somehow becomes unstable during, or following, prolonged rain and then undergoes a major readjustment before stability is restored.

Many landslides occur in periods of heavy rain. There is an increase in shear stresses, caused by an increase in total density (the increased soil water content in previously unsaturated material), coupled with the lower effective stresses and therefore lower frictional resistance due to increased porewater pressure. But this 'explanation' is incomplete: often it can be shown that equally wet conditions have occurred in the past without effect. We are therefore concerned with certain gradual changes which have proceeded far enough for the additional effect of, for example, exceptional rain, to initiate the movement.

Another common 'triggering' event is the oversteepening of a slope, by river or marine erosion, or other effects such that the shear stresses are increased to the point of failure. Such oversteepening does not necessarily involve any change of significance in the properties of the materials composing the slope, above the oversteepened part. On the other hand the exposure of sub-surface material to the atmosphere, as a result of mass movements, almost always initiates a series of changes in the strength properties of cohesive materials. An interesting illustration of this, is the failures which have occurred in railway cuttings some hundred years or more after their construction. The persistence of stable conditions for that period, suggests that the failures must be the result of long term changes in the materials themselves.

Reference has been made to the concentration of 'natural' landslides within particular localities. The juxtaposition of landslide scars gives a characteristic appearance to the topography, especially when viewed from the air, which is referred to as landslide topography. Examples are parts of southern Scandinavia, the eastern St. Lawrence region in Canada, in both cases post-glacial marine clays are involved, and in parts of the Dakotas, Montana, and Prairie provinces of Canada, where mainly tertiary materials are involved. The landslides occur most often during a spring snow-melt period or during heavy rains. But these can hardly be the ultimate causes, as they occur equally in areas of similar topography, which however are not so prone to landsliding. Detailed studies, prompted by geotechnical consideration, usually reveal long-term changes in the materials as an important factor, coupled with the effects of river erosion. Examples of landslide topography are shown in Fig. 9.4 and Pl. 9.1 and discussed further in section 9.8.

9.7 Loss of cohesive strength consequent upon removal of overburden

The shear strength of an intact rock sample is of course very high; it is largely cohesive strength as is illustrated in Table 9.1, even though the angle of internal friction may be much greater than in most soils. Consequently, a rock slope with even a vertical or near

Table 9.1 Shear strength of intact rock samples, c = cohesion, ϕ = angle of internal friction

Test: direct shear (D) or triaxial (T), and range of normal stress or cell pressure in 10^5 N m^{-2}.

Material	c, 10^5 N m^{-2}	ϕ °	Test
Chalk	9	21	T 0–70
Chislet siltstone	210	29	T 0–350
Pennant sandstone	350	44	T 0–350
Sandstone	42–420	48–50	T 210–2100
Limestone	35–350	37–58	T 210–2100
Granite	97–406	51–58	T 700–2800
Basalt-sandstone contact	10	59	D 0–20

Adapted from Carson and Kirkby (1972).

vertical face, might be expected to be stable even though of great height. A slope two thousand metres or more in height would not involve shear stresses equal to the strength of some intact rocks. Yet such very high vertical rock faces do not seem to occur. As Carson and Kirkby (1972) point out, the most likely reason is that the rock *in situ* does not in fact possess such high shear strengths. The reason is the occurrence of jointing, fissuring and cracking, which collectively are to be regarded as weathering processes. Frequently fissuring occurs because of the release of confining stress associated with denudation and the presence at the surface of the earth, of rock previously buried under great depths of overburden. When the confining stress is removed there is a tendency for irregular expansion. The cracking and fissuring process is often promoted too, by changes in temperature and moisture conditions of the kind associated with weathering in the narrower sense.

In a fissure, or crack, cohesion is absent. Areas of physical contact often occur of course, in a crack or fissure, but as fissuring proceeds, the stresses are increasingly borne by a limited amount of the cross-sectional area of the material. There is 'stress concentration'. Furthermore these concentrated stresses are ultimately resisted only by frictional strength, as the fissuring becomes more pervasive. The mobilizable strength in the material is far below that measured in the laboratory on an intact rock sample, and the maximum stable height for a slope is therefore much reduced.

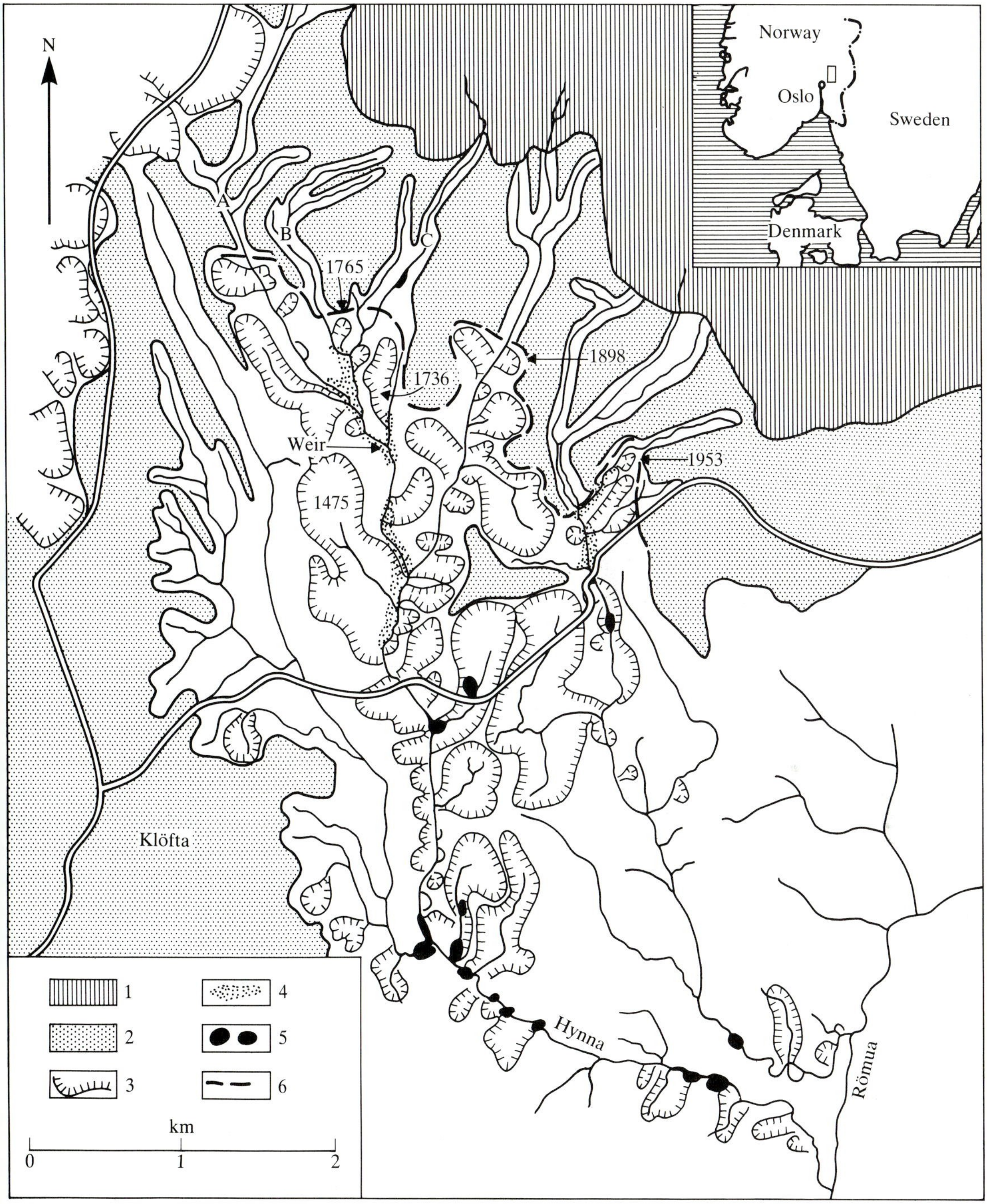

Fig. 9.4 Map of landslide topography, south Norway.
1. Bedrock area;   2. Old sea bottom plain;   3. Quick clay slide scars (dates of slides shown);   4. Re-deposited quick
clay slide masses;   5. Rock exposure;   6. 'Front of aggression'. (After Bjerrum, Löken, *et al.* 1969)

Plate 9.1 A landslide which occurred about 1000 years ago, in a river terrace of sensitive clay. As the soil mass moved the clay became liquid, as is evidenced by the flat form of the mass which has flowed out of the landslide scar area between the terrace front and the highway.

The concept of a material undergoing sufficient cracking and fissuring that it becomes essentially a granular, non-cohesive material also applies to clays. An early and classic analysis along these lines is that by Skempton and DeLory (1957) who drew attention to the predominance of slopes in the London clay having the angle *i* where $\tan i \simeq \dfrac{\overline{\tan \phi}}{2}$. This relationship suggests the case [9.10] of slopes of non-cohesive material with the water table at the surface. The inference is that the clay is so fissured as to have lost its cohesion, and during exceptional rains the water table comes to the ground surface. Numerous small displacements then occur to produce the slopes of angle *i*. The fissuring is of course, the result of drying and wetting, and other weathering processes. The cracked and broken nature of clay near the ground surface can often be observed in exposures.

A similar effect is observed in post-glacial marine clays (the Leda clay) of eastern Canada where a nodular structure of fist-sized pieces develops (Eden 1975). In this case, the material behaves as non-cohesive when the normal stress σ (weight of overlying material etc.) is small. At higher normal stresses, however, deformation does not seem to involve the nodules moving as units one in relation to another, and there is a significant cohesive strength.

In Chapter 4, attention was drawn to the decline in shear strength, from a peak value to a residual value, which is exhibited in laboratory shear tests with clays. The residual value is the resistance to shearing after a shear plane has developed. The plane represents a discontinuity across which, it is reasonable to suppose, there are no bonds to give cohesion. Many analyses of slides in fissured clays have shown that at the time of the slide the strength was, in fact, approaching the residual strength value (Skempton 1964).

The failure of slopes in clay or shale materials, after long periods of stability, is by no means limited to materials in which fissuring develops. Overconsolidated clays and shales, which have earlier been consolidated under effective stresses much greater than those currently existing, undergo various changes associated with their tendency to expand (Fig. 4.8). The most obvious effect is the loss of strength due to decreasing cohesion between particles with the decreasing density, that is increasing distance between particles. The water content simultaneously increases. If these effects occur unevenly they may also promote fissuring.

Depending on the degree of consolidation shale materials at depth may be quite hard and rocklike ('well-bonded'). While these break down on unloading, into numerous more or less discrete pieces, 'poorly bonded' materials may absorb enough water on unloading to become a soft clay (Terzaghi and Peck 1967). In either case the loss of strength makes

landsliding more likely, as the unloading process continues. Such weakened materials are likely to occur in river valley sides, these representing situations where greater removal of overlying material has occurred.

Bjerrum (1968) introduced the concept of residual strain energy in relation to '*diagenetic bonding*'. He considers the process of consolidation in large measure to involve bending of flaky clay particles under load. When the load is removed, the particles' tendency to straighten is responsible for much of the volume increase (or rebound). The strain energy may however, remain 'locked in'. This is the case if, after consolidation, cohesive forces, cementation or other effects constituting diagenetic bonding serve to give some rigidity. Such bonding is eventually destroyed by weathering processes so that much of the rebound occurs some time after the unloading. Bjerrum also emphasizes the increase in lateral stress that occurs in association with the expansions. Such stresses give an added component of shear stress in an essentially downslope direction, simultaneously with the reduction in strength.

Associated with most of the changes in soil structure described in this section, there is often pronounced creep, which accelerates prior to a landslide occurring. Monitoring of creep movements in slopes believed to be unstable is valuable in the prediction of slides.

9.8 Sensitivity

The chemistry of soils is important for their strength in several respects. Cohesion in clay soils is very significantly influenced by the cations present in the pore water. The effects arise largely because of the positive electrical charge of the cations, which may accordingly attach themselves to the negatively charged clay surface, and to the negative dipole of the water molecules. Thus the cations may be visualized as providing a bonding mechanism between adjacent clay particles, with or without the intervention of intervening water molecules.

The relationships are in fact very complex, and are the subject of a large literature (see e.g. Mitchell 1976). They require a deep knowledge of surface chemistry, and of adsorption theories. While these fundamental aspects are beyond our scope in the present context, some effects will be briefly reviewed.

In an early work, Rosenquist showed that landslides in post-glacial marine clays, involving a near-liquefaction, were associated with the leaching out of marine salts (Bjerrum 1955). The clays were 'extra-sensitive' or 'quick'. Such clays are characterized by their assuming a much lower strength, on disturbance. One is reminded of the difference between peak strength and residual strength, except that in the sensitive clays the strength may fall to a small part of the 'undisturbed' value. Sensitivity is defined as:

peak undisturbed shear strength
remoulded (or 'disturbed') shear strength

In extreme cases the ratio may be a 100 or more. Thus in the disturbed state much material is essentially liquid, and will give flows rather than slides. Such flows have been responsible for much loss of life and property damage, particularly in Norway, where research has been extensive (see review by Rosenquist 1966). If sufficient material liquifies, even the slightest slopes will fail. There are also many slides which occur because of particular, perhaps isolated, extra-sensitive layers, amongst otherwise more secure materials.

Almost all clays exhibit some degree of sensitivity in so far as there is some loss of strength on disturbance, but if we direct attention to those of sensitivity perhaps 15 or more, we identify materials where failure is not restricted to a well-defined shear plane. Rather, once disturbance is initiated, the effect spreads: the soil structure is such that it collapses throughout a considerable body of soil which then has uniformly low strength. Such structure has been likened to a 'house' of playing cards.

Many theories have been advanced to explain sensitivity (Mitchell and Houston 1969). As early as 1932 Casagrande postulated that consolidation of clays containing silt and sand would result in the larger particles forming a load-bearing framework, while within the openings in this framework clay would be relatively protected from consolidation. If the whole were then mixed, more of the load would fall on to the little-consolidated clay which would be correspondingly low in strength. Later theories, in addition to the simple leaching concept explored by Bjerrum and Rosenquist (1956) have included a loss of cementing agents (by weathering) to leave an unstable particle assemblage, and complex cation exchanges and substitutions over long periods. It must also be remembered that many materials, butter, some paints, and ketchup as examples, are thixotropic, that is, they flow, or at least are weakened, on being disturbed. Although thixotropy is also found in clays, under field conditions sensitive clays obviously do not return to their original structure after liquefaction. Sensitivity is distinguished from thixotropy in this manner. Ultimately, disturbed sensitive clays regain strength but this involves changes of porewater pressure, and re-establishment of bonds between particles, over a prolonged period. Furthermore, the new bonds are of a somewhat different nature to those previously existing. The structure prior to disturbance, is after all, the product of thousands, even millions of years of physical and chemical modification, of a sedimented material.

Quick clays in Eastern Canada ('Leda' clays) unlike those in Scandinavia, do not show a simple correlation between decreasing salinity and sensitivity, although otherwise rather similar. Torrance (1975) demonstrated that this is not because salinity and cation composition is unimportant. It results from the rather variable nature of the clays in respect to other factors which also affect strength. The Leda clays and their Scandinavian counterparts are composed mainly of illite and chlorite. Clays composed of montmorillonite behave quite differently, and increasing cation concentration may be associated with *loss* of strength.

The full understanding of quick clays, and as a result, a minimizing of the threat they pose to man, is some way into the future. The quick clays illustrate how important apparently insignificant material characteristics may be, in relation to major phenomena such as landslides. The concentration of salts, to which so much attention has been given, are often measured in a few grams per litre, that is, insufficient to give significant taste to the pore water. Sea water has a concentration of about 35 g ℓ^{-1}, but it is only when the (originally marine) pore water solution has been reduced to about 5 g ℓ^{-1} that sensitivity may increase rapidly. One reason for the quick clay occurring in pockets of limited extent is the tendency for an exponential increase in sensitivity at very low concentrations.

In general the quick condition is itself only temporary and further weathering may involve an increase in cations which decreases the sensitivity again. The type of cation involved is important, as well as concentration. The rates of water flow causing leaching are likely to be greater where the clay layer (which has low permeability) is thinnest. Figure 9.5 illustrates the location of a quick clay mass above a hill in the bedrock. There is firmer clay below the quick material. In this example the leaching involves a flow *upward* of water coming from the bedrock (see also Moum and Torrance 1971), and the lower, less sensitive material,

has earlier experienced the 'quick' stage. The additional cations are magnesium, and probably result from a disintegration of the clay mineral chlorite. The cations largely responsible for the greater strength of the surface weathered layer, are, by contrast, mainly aluminium and iron, released by oxidation associated with infiltration of surface water.

Löken (1970) showed the effects of adding various cations to a quick clay; in all cases there was an increase in the liquid limit and very variable increases in shear strength. The liquid limit (a frequently measured quantity in soil engineering) is simply the lowest water content at which a soil liquifies on being subject to a standardized mechanical disturbance. A clay which is quick necessarily has a water content above its liquid limit. The effect of an increase in liquid limit, to above the *in situ* water content, is therefore loss of sensitivity. The increases both in liquid limit and shear strength observed by Löken were dependent on the species of cation, and sodium (Na^+) was least effective, while aluminium (Al^{3+}) was most effective. Of course, sodium is the most abundant ion in the sea water, and in this respect its removal is significant. But Löken's work demonstrates that detailed chemical studies are essential to explain correctly the effects of weathering and leaching, subsequent to removal of the original

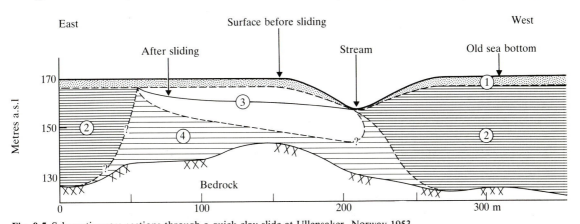

Fig. 9.5 Schematic cross sections through a quick clay slide at Ullensaker, Norway 1953.
1. Weathered crust;   2. Marine clay;   3. Marine clay, leached until quick;   4. Marine clay leached until quick then weathered. (After Bjerrum, Löken *et al.* 1969)

marine salt. Finally, it should be noted that the addition of anions, CO_3^{2-}, OH^-, PO_4^{3-} etc. from various salts tends to promote weakening of the clay, because these are dispersing agents, having roughly the reverse effect to that of cations (Torrance 1975).

9.9 Evolution of slopes

Study by geomorphologists of the sequential development of slopes preceded the realization of the significance of the mechanical properties of earth materials by many decades. These early studies were often excessively theoretical and academic, and frequently involved misconceptions. The strength of the constituent earth material is both a cause and a result of the manner of development of a slope. Kenney (1967) demonstrates the relation of consolidation, weathering and thus soil strength, to sea level changes and late and post-glacial denudation.

Most natural slopes differ greatly from the 'ideal'. Not only do the materials usually change, in a geological sense, within the limits of a 'single' slope; but also it is almost inevitable that the depth to water table, and the pattern of natural drainage, must be different in different parts of a slope. Consequently, any mathematical model treating a slope as a body of uniform material is open to suspicion, a suspicion which is strengthened by the complexity of the slope forms we see around us.

On the other hand analysis of the sequence of events through time, on particular slopes whose nature is well-defined, aids in assessing the changes which lead to major mass movements. Changes in the Canadian 'Leda' clay as a result of near-surface exposure to weathering have been referred to several times (pp. 139–41). Active stream erosion of this material is common, and each spring bare slip surfaces, several metres high, appear, as segments fall into swollen streams. These slip surfaces are steep, often almost vertical,

because of cohesion. During the summers, drying and thus fissuring occurs, with further, smaller, movements reducing the steepness of the slope. As the valley enlarges with higher slopes, the slips due to the stream erosion are restricted to the lower part of the slopes (Fig. 9.6). The upper parts of the slopes become grass covered, and often show small step-like terracette forms. It is suggested by Carson and Kirkby (1972) that the terracettes (which occur quite commonly on clay or sandy-clay materials) arise because the surface layer is strengthened by roots, while the underlying material is essentially unstable. This idea is supported by observations by Mitchell and Eden (1972) showing creep to occur to depths of 10 m (Fig. 9.7). The creep movements do not necessarily lead, at least in the short term, to landslides, although these are common and often involve the entire height of the slope. The landslides represent a further, abrupt, stage in the development of slopes in the 'Leda' clay. Indeed as Eden (1975) points out, such landslides probably follow from further development of the fissuring and attendant weakening (section 9.6). The 'Leda' clay being, on occasion, highly sensitive, a failure of the flow type is also a possibility as a further step.

In this example the evolution of the slope is associated with a series of changes in the material, each leading to lower strength and correspondingly, lower slope angles. It is not necessarily the case that such changes lead to lower strength. In fact, in many clays the near-surface layers are stronger than deeper-lying layers. Drying during the summer months leads to large suctions, and effective stresses, with consolidation and increased strength (Chapters 4 and 6). Even on rewetting, increased strength persists due to the over-consolidation. Freezing has similar effects (Chapter 11 p. 177), which occur to great depth in areas of former permafrost. In addition to such physical changes, which have led to the term 'drying crust', chemical changes to a depth of many metres may also be responsible for a strength increment as noted earlier.

The existence of such a stronger, less-

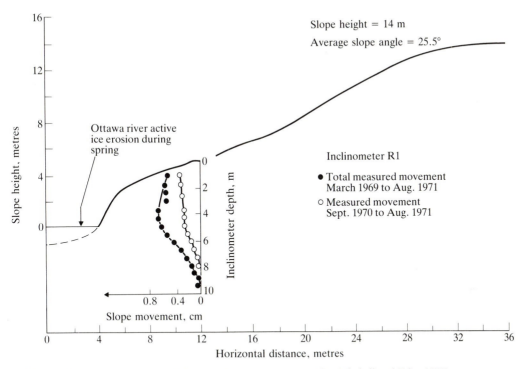

Fig. 9.6 Measured movement in a natural slope of clay, near Ottawa. (After Mitchell and Eden 1972)

sensitive weathered layer is significant in the evolution of slopes in the slide-prone quick clays in Norway. According to Bjerrum *et al.* (1969) the location of major quick clay slides is related to active stream erosion (Fig. 9.4). Initially, the stronger 'drying crust' allows the development of relatively steep-sided valleys. The thickness of the crust depends on the rate of its formation, compared to that of its removal by shallow movements (10–20 cm) of a surface fissured layer. The fissuring due to drying, reduces the strength of the weathered material to that due solely to friction; the material is then prone to displacements during wet periods. If a thick strong crust develops, a steep slope of considerable height can form as the stream cuts down, only to suddenly fail, releasing the weaker, sensitive material behind it.

A similar stage of landslides following the development of a valley side to a critical height (cf. Fig. 9.1), by relatively rapid downward erosion of a river, is envisaged by Skempton (1953). Initially there are shallow surface slides

and slumps, but when the valley side becomes high enough major deep-seated slides occur. Subsequently, once the river erosion ceases, the slopes tend to decline slowly to a more or less uniform angle, again by shallow movements, no doubt in association with the infrequently occurring maximum pore water pressures (see sections 9.3, 9.7). This, and the previous examples, illustrate the complex manner in which changing soil properties and changing slope geometry, influence the course of development as well as the present stability of slopes.

9.10 Overland flows and slope evolution

The foregoing discussion has directed attention to mass movements or displacements of material, in which some depth of ground is involved and where circumstances have become such that the shear stresses exceed the shear strength. In addition movements of material over the *surface* of the ground occur, due to forces of gravitational origin made effective by running water, or forces associated with wind. The nature of the soil and of the surface also have a controlling influence on this kind of movement. We restrict our considerations to ephemeral water flows over the ground surface.

When such displacements affect the quality of agricultural land, they constitute soil erosion in the agricultural sense, and have immediate economic importance. Some localities are renowned for such problems but the phenomenon is in fact very widespread. Observations in an experiment in Ottawa (not a locality regarded as particularly susceptible) showed a loss of 15 cm of topsoil in 59 years (cited in Williams 1963). Downslope movements of surface particles occur to some degree in all kinds of soils and comminuted rock, even though the amounts may be small. They assume importance to the geologist and geomorphologist as a land-forming process. Recent research has drawn attention to the importance of the processes in various aspects of environmental management, including for example river and reservoir sedimentation, and the role of surface characteristics in this regard (Heinemann and Piest 1975).

For significant water erosion to occur, the soil must be of low enough permeability that run-off occurs. The rate of infiltration must be less than that of precipitation and surface flow (that is, flow from the adjacent surface, upslope), combined. The pre-existing moisture contents, and soil drainage is therefore also important. Particles within certain size limits are dislodged by raindrop impact. The greater part of the displacement of material follows from transport in flowing water. A tractive force, essentially due to the weight of the water, is exerted on the ground surface (the expression for which is similar to that for shear stress in the ground [2.1]), while particle movement is resisted by the weight of larger particles, and by cohesion associated with small particles.

The water may flow over the surface as a sheet, or it may have a strong tendency to become concentrated into small, eroded channels a cm or so in width and known as rills. These may also enlarge to give gullying. Such effects, which are also a function of soil type, are considered in texts on fluvial geomorphology such as Leopold, Wolman and Miller (1964).

Both the occurrence of over-surface water flow, and its erosive effect consequently depend on many factors relating to the composition of the soil, the geometry of the surface, including eventual vegetation, as well as the precipitation pattern. This had led to empirical methods in which water erosion is predicted by expressions of the type (Beasley 1972):

$$A = RKLSCP \qquad [9.11]$$

where A is mass eroded, per unit area and unit time

R is rainfall factor
K is soil erodibility factor
L is slope length factor
S is slope gradient factor
C is crop management factor
P is erosion control practice factor

In practice the value to be ascribed to each factor is obtained from charts or tables. These show the values corresponding to a range of possible site characteristics relative to the factor. Such information is derived from test plots with specified conditions, and ideally, in which the effect of a single variable can be isolated.

Another related form of erosion is internal erosion of the soil by permeating water. A phenomenon known as piping which is of importance particularly in dams, involves the movement of particles small enough to be carried through the pore structure (Terzaghi and Peck 1967). More widespread is the solution of soil constituents and their removal in soil water and subsequently rivers. Observations by Rapp (1960), although open to some questions of interpretation, suggest that removal of material in this way is a major component of the overall denudation in Arctic-Alpine climates.

Remarkably, the importance of oversurface movement, relative to that of deeper mass movements, in the formation of slopes, has not been subjected to much scientific investigation. Too frequently, there is a tacit assumption of the dominance of the one over the other, while Leopold, Wolman and Miller (1964) in an interesting discussion conclude that most slopes are the product of *both* processes. Nevertheless, the present form of the slope is often ascribable to one much more than the other, and the distinction can be very important.

Firstly, and fortunately, most slopes are stable – that is, they have a safety factor with respect to landslides or similar rapid movements (involving a depth of soil), substantially greater than one. Furthermore, radically different circumstances and conditions, so different that we could scarcely regard the slopes as being the same slopes, would often be required for this not to be the case. Therefore, there must have been a quite long interval, geologically speaking, during which such slopes have been evolving to their present form by oversurface displacements, very shallow movements or a slow creep which may extend to a significant depth. Such creep (Kojan 1967) should be distinguished from that associated with the processes leading to sudden failure (as discussed in relation to overconsolidated clays and shales).

Secondly, where unstable slopes occur frequently ('landslide topography'), that is, where significant, rapid, mass movements have occurred recently, it is those which give the characteristic expression to the slope form. The strictly surface movements or continuous, non-accelerating creep are then relatively unimportant in that respect.

10

Climate below the surface

10.1 Soil climate

The thermal behaviour of soils in general was considered in Chapter 5. In this chapter the climate within the ground is considered. This involves the temperatures and heat flows occurring at different depths and times. It also involves the soil moisture potentials and moisture flows. We are considering a layer of the ground of limited depth, and the climate in this layer follows from atmospheric and astronomic conditions, just as does climate in the more general sense. A broad approach is often taken to 'soil climate' where the mass and energy exchanges at the ground surface would be considered in detail under that heading. They were considered in Chapter 8. It is convenient to consider a plane immediately underlying the surface as the *boundary of the ground* for the purposes of this chapter. This is because, once the temperature at that plane, or alternatively, the heat flows through it, are defined it becomes possible to examine the conditions in the ground without further regard to the complex and varied surface energy exchange processes. This applies too, to soil moisture.

The soil thermal regime is primarily a matter of conductive heat flow. Radiation occurs only across pores which are empty, and only exceptionally does such radiation need to be considered. Evaporation and condensation are important elements in heat transfer in certain soil moisture content ranges. Transfer of mass in vapour or liquid form in the presence of temperature gradients necessarily involves heat transfer. If the thermal conductivity coefficient λ for the soil in bulk is determined, these effects are accounted for, even though physically they are processes distinct from thermal conduction in the strict sense. Complicated approaches involving analysis of heat and mass transfers as coupled (i.e. interdependent) phenomena, are not yet fully developed and lie outside our scope. We shall therefore consider thermal and moisture conditions separately.

10.2 Cyclic temperature change

Just as the temperature of the air varies in a cyclic manner both annually and diurnally, so does the temperature at the boundary of the ground. It is not correct to regard the one temperature as the cause of the other. Although the air and ground temperatures may in some situations be closely similar, differences between them are often significant. Both are determined by the various energy exchange processes occurring at and in the vicinity of the ground surface. Generally the range of temperature within the soil is less because of the barrier to heat exchange represented by surface cover, vegetation, snow etc., and the overlying soil itself. There are exceptions, for example as when solar radiation falls on dark bare soil or rock. Numerous short-term variations are characteristic of the air temperature, but not of the soil below the surface.

If air temperatures are plotted as monthly means, a rather regular curve (a sine wave – so-called because it can be represented by an equation involving a sine function) is obtained for the annual temperature wave (Fig. 10.1). A similar wave, generally of smaller amplitude, describes the temperature at the ground boundary.

Temperature change is transmitted into the ground by flow of heat inwards and outwards. Such heat flows occur continually along changing temperature gradients, as in the illustrations (Fig. 10.2a and b). Change of temperature at the ground boundary is followed by change within the ground after a time interval. The magnitude of the change decreases with depth. For example the summer maximum is detachable at some 4–6 m depth in mid-winter, but is much less than at the ground boundary.

The effect of imposition of a cyclic temperature wave on a uniform solid body is described by equations of heat flow derived many years ago, and subsequently applied to soils (Keränen 1929). The equations are

— Mean annual temperature

⌢ Sine wave

o Mean monthly temperature

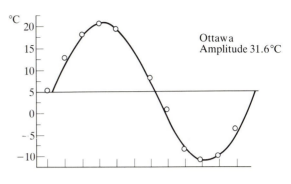

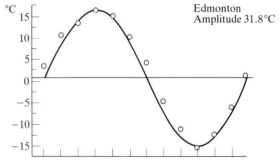

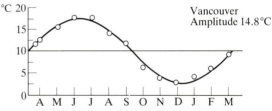

Fig. 10.1 Mean air temperatures, by month plotted for the whole year. (After Williams and Nickling 1972)

discussed in detail in for example Grober, Erk and Grigull (1961). Although the conditions under which the equations would be exact are not met in soil the following simple relationships are of particular value in interpreting ground temperatures (Penrod, Walton and Terrell 1958).

If the amplitude at the boundary is S_b °C, then the amplitude S_z at depth Z below is given by:

$$S_z = S_b e^{-Z\sqrt{\frac{\pi}{at}}}$$

[10.1]

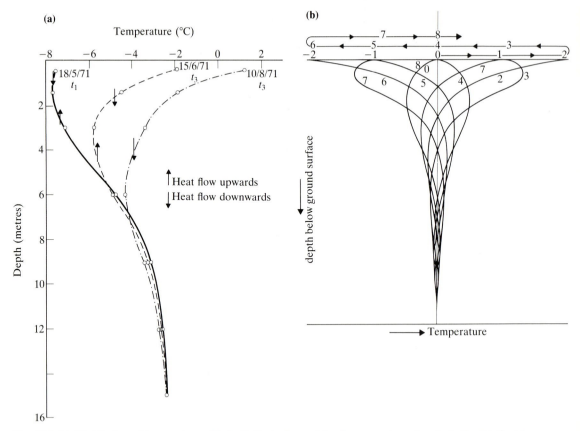

Fig. 10.2 (a) Distribution of temperature with depth (tautochrones) for three occasions at a site in the Northwest Territories, Canada. (Modified from Smith 1976)
(b) Ideal temperature – depth curves (tautochrones) for eight different times, presuming the surface temperature to be rhythmically varying between a maximum and a minimum value corresponding to the succession of day and night or summer and winter temperatures. (After Beskow 1935)

or $\dfrac{\ln S_b}{\ln S_z} = -Z\sqrt{\dfrac{\pi}{at}}$

where a = diffusivity m² s⁻¹
       t = period, i.e. 1 yr = 3.1536 10⁷ S
       e = exponential

In fig. 10.3 maximum and minimum annual temperatures are shown as a function of depth. Note that although the value of diffusivity has a marked effect on the amplitude, differences in climate, between say a maritime, or continental, location may be more important.

The delay Δt between the occurrence of a certain point on the temperature wave (for example the summer maximum) at the boundary, and at depth Z below the boundary is given by:

$$\Delta t = \frac{Z}{2}\sqrt{\frac{t}{a\pi}} \qquad [10.2]$$

A further equation:

$$Z_\ell = 2\sqrt{\pi a t} \qquad [10.3]$$

gives approximately the maximum depth Z_ℓ at which a cyclic fluctuation will be felt. Because there is no temperature term, it appears that the annual wave will be felt to a similar depth regardless of whether the amplitude at the boundary is large or small. In fact, because of the manner in which the amplitude decreases with depth, *perceptible* change (for example 0.1 °C) will be restricted to shallower depths where the amplitude at the boundary is smaller, as illustrated in Fig. 10.3. Somewhat more

Temperature °C

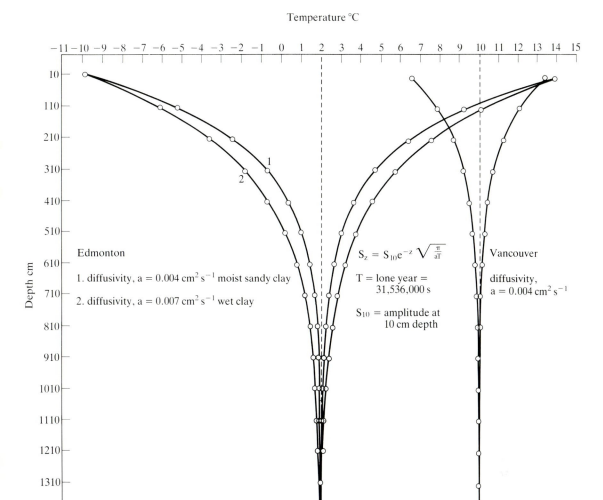

Fig. 10.3 Calculated annual amplitudes S_z of temperature (After Williams and Nickling 1972)

complicated equations allow calculation of temperatures as a function of time and depth, for specified boundary temperatures and soil diffusivity (van Wijk 1966). The equations have also been extended to cover the case of soil layers of differing diffusivity.

It is important not to underestimate the deviations from ideal behaviour that occur because of the complexity of natural soils and soils profiles. Nevertheless, the equations may often be applied to observations of ground temperatures through the year at various

depths and times, to obtain a value of diffusivity. For example, if Δt in [10.2] is plotted against $\dfrac{Z}{2}$ the slope of the line gives $\sqrt{\dfrac{t}{a\pi}}$. Such calculated values may be more reliable than those obtained by laboratory procedures. It is desirable to make further calculations using other equations, and to ascertain that the values are consistent. If they are not, the most likely reason is a non-uniform soil.

Variations in thermal properties during the

course of the year produce some asymmetry in the pattern of temperature change. The most important effects are summer drying of the upper layers and seasonal fluctuations of water table (Fig. 10.4), and winter freezing. The large quantities of latent heat released during freezing have the effect of retarding the penetration of the 0 °C isotherm A similar circumstance occurs during the spring thaw. The rates of temperature change at temperatures just below 0 °C are also greatly reduced because they involve exchange of large quantities of latent heat associated with changes in the proportion of ice and water (Chapter 7). To the extent that the temperature change in the boundary does not follow a sine wave form, for example under winter snow cover, the passage of temperature in the ground will also be affected.

The diurnal passage of temperature is usually not well represented by a sine wave, in contrast to the situation with the annual wave based on monthly mean temperatures Fig. 10.1. Weather conditions may be such that night time temperatures vary little from daytime, or periods of sunshine may produce large irregular peaks. Thus application of equations of the form of 10.1, 10.2 and 10.3 is of limited value, although providing a useful idealization.

Equation 10.3 shows that the maximum depth at which a perceptible temperature change occurs, depends on the square root of the period of the fluctuation. Thus, if for a given soil an annual wave is felt to a depth of 10 m, then in the same material a diurnal wave would be felt to $10 \times \dfrac{1}{\sqrt{365}} = 0.52$ m. Even at depths less than this there will be a marked

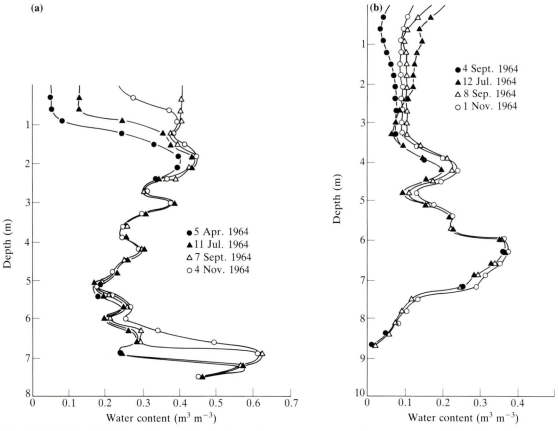

Fig. 10.4 Soil water content profiles at different times of year, two sand soils, forest covers. (a) is taken from Young sand, Maint Gambier Forest and (b) is taken from Kalangadoo sand, Penola Forest. (From Baver, Gardner and Gardner 1972)

reduction in the amplitude, in a similar manner to that illustrated in Fig. 10.3. Not infrequently temperature measurements are reported which appear quite contrary to the implications of [10.1], [10.2] and [10.3]. There are a few natural circumstances, in which such observations can be realistic. For example a crack extending from the surface may allow almost immediate transfer of heat by percolating water from the surface such that temperatures of equal magnitude are observed almost simultaneously at the surface and at depth. But in the majority of cases, such 'anomalous' readings are due to instrumental error. Measurement of ground temperatures is quite difficult and rarely achieved with an accuracy better than a tenth of a °C. The pattern of temperature change is one of increasing stability and regularity with increasing depth. This is important in considerations of weathering, and for example, in the distribution of permafrost. Cycles of freezing and thawing in particular are greatly reduced in number at only a few cm depth.

10.3 Mean annual ground temperature and geothermal gradient

So far, we have considered the transmission of temperature changes into the ground. A separate question is that of the temperature level of the ground and its relation to location. The temperature level is conveniently considered in terms of the mean annual ground temperature. The latter can be measured directly at any time in the ground below the depth reached by the annual variation. A similar value is obtained for shallower depths by appropriate averaging for a year, of the temperatures occurring periodically, for example monthly. In the latter case, the depth of observation must obviously be below that reached by diurnal fluctuations.

The mean annual ground temperature commonly differs, even by as much as 10 °C,

from the mean annual air temperature. In this respect there is a temperature discontinuity at the earth's surface, which is not surprising in that the heat exchange is not only a matter of simple conduction. It is the complex and various processes of energy and mass exchange at the ground surface which are responsible for the magnitude of the difference in mean annual temperatures between ground and air. The mean annual temperature below a paved road is usually different by a few degrees from that below adjacent grass-covered areas, and both usually differ from the mean annual air temperature. The latter is generally more uniform laterally, and at least it does not vary in a similar manner. The mean ground temperature is only indicated in a general way by mean annual air temperatures. This has important practical implications for agronomy, plant ecology, and particularly in frozen ground engineering (see p. 163).

In regions of substantial snow cover, the mean ground temperature is normally higher than that of the air, because the ground is shielded from winter cold. But in the north-west Norwegian coastal districts for example, the reverse is reported. This is probably a result of the low net incoming radiation, the sparse snow cover and the relative warmth of the moist air (associated with the gulf stream).

The uppermost soil layers may have a slightly different mean ground temperature from deeper layers, due to different thermal properties, particularly during freezing and thawing. These gradients of mean temperature do not necessarily imply a state of long-term thermal disequilibrium – a slow warming or cooling of the ground. The different thermal properties at different times of the year in fact require slight mean annual temperature 'gradients' if a quasi-equilibrium state is to persist; that is, if the annual intake and loss of heat are to equalise.

A change in ground surface conditions, such as removal of vegetation cover, or a change in its composition will however initiate changes of mean ground temperature which are transmitted slowly to greater depths at a rate proportional to the square root of the time since the change occurred. This 'square root of time' relationship (also indicated by [10.3]) is useful in providing an approximation of the time interval before surface changes, whether climatically or microclimatically induced will be reflected in temperature change at particular depths.

The temperature below the maximum depth of annual variation, increases at a rate of approximately 1 °C per 40 m. This constitutes the *geothermal* gradient, the gradual increase in temperature being due to the heat of the earth's interior. Its value varies somewhat in different parts of the earth (Lee and Clark 1966). The geothermal gradient is also not likely to be uniform with depth. It changes with the conductivity of the rock or soil. If it did not, of course, the heat flow would not be constant. There would be a continual accumulation or loss of heat with attendant temperature change in particular layers.

Heat is in fact accumulated or lost from certain layers due to past climatic change (Cermak 1971). In ground subject to different temperatures in association with the quaternary glaciations there may be maxima or minima in the geothermal gradients depending on the duration of, and at depths depending on the square root of time since, the different boundary temperatures. The ground surface under the ice sheets in some cases was warmer than current surface temperatures, while around the borders it was lower. Other sufficiently long-term climatic fluctuations are also 'recorded' in disturbances of the geothermal gradient, and consequently in the geothermal heat flow. The analysis of such disturbances is difficult however, and may not provide much information relative to past climates.

10.4 Heat flows in the ground

Figure 10.2a illustrates the fact that at any time heat is flowing upwards, and downwards, the direction of the temperature gradients changing with depth. The magnitude of the heat flows may be calculated rather simply from repeated observations of tautochrones (Fig. 10.2a and b) together with values of soil heat capacity. The product of the heat capacity C of a thin layer ΔZ, and the temperature change ΔT, in a time interval, that is $\Delta Z\, C\, \Delta T$, gives the heat gained or lost from the layer. This passes through the upper or lower horizontal surface of the layer and the quantity per unit time, and per unit area of the horizontal surface is:

$$Q = \frac{\Delta Z\, C\, \Delta T}{\Delta t}$$

$$\Delta t = \text{time interval, } s$$

with the units:

$$\frac{\text{m J m}^{-3}\ {}^{\circ}\text{C}^{-1}\ {}^{\circ}\text{C}}{\text{s}} = \text{J m}^{-2}\ \text{s}^{-1} = \text{W m}^{-2}$$

The total heat flow Q through unit area of a horizontal plane is given by the sum of the heat lost or gained by all n layers below:

$$Q = \frac{\Delta Z_1\, C_1\, \Delta T_1 + \Delta Z_2\, C_2\, \Delta T_2 \ldots \ldots \Delta Z_n\, C_n\, \Delta T_n}{\Delta t}\ \text{W m}^{-2} \quad [10.5]$$

Terms for those layers absorbing heat and therefore having a rising temperature are positive, and upward heat flow therefore has a negative sign.

An even simpler procedure is to calculate the heat flow through a layer at the horizontal plane of interest, from the product of thermal conductivity and the temperature gradient through the layer: $\lambda\dfrac{\Delta T}{\Delta Z}$. This procedure suffers from inaccuracy due to the difficulty of assessing λ (see p. 62).

Typical values for heat flow during the winter months through planes 20 cm and 2 m below the surface of an asphalted road are given in Table 10.1. It is apparent that the upward heat flow due to the geothermal gradient, some 0.05 W m^{-2}, is quite insignificant compared to the heat flow originating from the storage of heat (heat intake) during the previous summer half-year. The figures for heat flow upwards through the 20 cm plane approach 25 W m^{-2} (table 10.1a) during December, for example, and would be substantially greater for the ground surface. The figures illustrate the possibilities for storage of solar energy during warm (summer) periods, for use during cold winter periods.

Tables 10.1a and 10.1b show heat flows for two different soils under otherwise similar conditions. The annual exchange is represented by the sum of heat flows (at a particular depth) of the same sign. Although the data are somewhat incomplete it is clear that the annual exchange through the surface of clay ground is less than that of the sandy-clay ground. The thermal conductivity of the sandy-clay was greater than that of the clay by a factor of about four, while its heat capacity was less by about 20 per cent. The conductive capacity coefficients of the two types of material are therefore roughly in the ratio 1 : 1.8. This agrees well with the ratio of observed heat exchange of about 1 : 1.9.

The conductive capacity coefficient $\sqrt{\lambda C}$ (p. 69) provides a good indication of the ability of the ground to absorb and release heat in response to specified ground surface temperatures. Rather few determinations of cyclical ground heat flows have been made, and thus of the annual exchange of heat of the ground, in relation to soil conditions. In recent years renewed attention has been given to the question, particularly in response to geotechnical concerns with penetration of frost, and to agricultural and other problems relating to storage of solar energy.

10.5 'Warm' soils and 'cold' soils

Implicit in the use of [10.1], [10.2], [10.3], is the assumption that the temperatures at the upper boundary of the ground are defined. Clearly this approach has limitations in that these temperatures must be determined as an element in a system involving all components

Table 10.1a Heat flows, through horizontal planes below asphalt highway, Ottawa, Canada.
Positive figures: heat flow downwards. Negative figures: flow upwards.
Clay subsoil, site 0.5 km distant from that for Table 10.1b.

	Heatflow (average over period of obsn)	
Period of observation	*at 20 cm depth* W m^{-2}	*at 2 m depth* W m^{-2}
Sept. 24 to Oct. 16	− 7.2	+ 0.8
Oct. 16 to Nov. 2	− 6.5	− 1.1
Nov. 2 to Nov. 17	−11.5	− 2.1
Nov. 17 to Dec. 4	−14.3	− 3.1
Dec. 4 to Dec. 14	−24.5	−13.1
Dec. 14 to Jan. 4	−3.2	−1.9
Jan. 4 to Jan. 19	− 5.6	− 2.1
Jan. 19 to Feb. 5	−12.0	− 6.3
Feb. 5 to Feb. 22	− 5.8	− 5.0
Feb. 22 to Mar. 7	− 6.3	− 5.8
Mar. 7 to Mar. 21	− 8.1	− 8.1
Mar. 21 to Apr. 7	− 0.2	− 1.1
Apr. 7 to Apr. 21	+14.4	− 2.8

From field observations for Ontario Ministry of Transportation and Communications, Williams (1975) Thus, total annual outward flow of heat at 2 m depth (calculated with corrections for number of days in half year) $\simeq$ 64 10^6 J m^{-2}. This figure represents annual exchange of heat for ground below 2 m, the annual outward flow being balanced by an approximately equal inward flow.

Table 10.1b Heat flow sandy-clay subsoil Ottawa W m^{-2}

	at 20 cm	*at 2 m*
Sept. 17 to Oct. 16	− 0.56	+ 6.5
Oct. 16 to Nov. 2	− 2.2	+ 2.3
Nov. 2 to Nov. 17	− 9.1	+ 0.82
Nov. 17 to Dec. 4	−12.4	− 1.0
Dec. 4 to Dec. 14	−18.4	−10.8
Dec. 14 to Jan. 4	− 8.6	− 5.9
Jan. 4 to Jan. 19	−12.4	− 8.9
Jan. 19 to Feb. 5	−15.7	− 9.1
Feb. 5 to Feb. 22	−10.9	−10.7
Feb. 22 to Mar. 3	− 8.3	−10.1
Mar. 3 to Mar. 21	−15.3	−10.5
Mar. 21 to Apr. 7	− 8.3	− 9.5
Apr. 7 to Apr. 21	+7.2	−13.7

Total annual outward flow of heat at 2 m depth (calculated with corrections for number of days in half year) $\simeq$ 121 10^6 J m^{-2} (compare value of 64 10^6 J m^{-2}, for clay subsoil).

of the energy exchange at the earth's surface.

The annual or diurnal range of temperature at the ground surface can vary greatly within any locality, even in the absence of any measurable difference in the gross climatic elements. Not only are the microclimatological properties of soil surfaces, such as the albedo, or the availability of soil moisture for evaporation, important as discussed in Chapter 8. In addition the temperature at the ground boundary is affected by the distribution of heat within the ground and its variation with time. These quantities depend on the thermal properties. Variations *in soils or surface conditions* generally control the heat flows at all depths subject to annual variation, to a greater extent than the above ground climatic parameters. Thus, there are the marked differences associated with soil variations, illustrated in Table 10.1a and b. One would have to travel hundreds of kilometres to find similar differences due to *atmospheric climatic differences* alone. Because variations in water content have a great affect on thermal conductivity and heat capacity, the hydrological or 'drainage' conditions of the ground are important together with the dependence of volumetric water content on grain size composition or organic content.

Taylor (1967) reports on observations at 20 cm depth, from adjacent peat and sand sites, which show that the sand has higher temperatures in the spring. Doubtless the vegetative covers at the sites were quite dissimilar, but the main reason is probably the lower conductivity of the sand, and its lower heat capacity. The lower conductivity retards the dispersal of heat downwards, while the lower heat capacity results in a rapid temperature rise in a shallow surface layer of the sand, a temperature rise which is large in relation to the amount of heat entering the ground surface. During the autumn the difference between the two sites is less marked, but there is a comparable effect. That is, the cooling of the peat is slowed as a consequence of its thermal properties.

Observations of this kind can also to some extent be understood in terms of [10.1] and [10.2]. Low conductivity can be the cause of low diffusivity $(\frac{\lambda}{C})$ values which (Fig. 10.3) are associated with rapid reduction of amplitude. A smaller value of diffusivity will also give a greater time lag in the arrival of a point on the annual wave [10.2]. The example quoted serves to illustrate a certain difficulty: the wet peat having higher conductivity than the drier sand might be expected to cool more rapidly in the autumn. In fact, the heat capacity of wet peat is greater than that of dry sand probably by a much greater factor, such that the peat actually has a lower diffusivity than the sand. This being the case, the observations are in accordance with [10.2].

There are published many facile interpretations of observed ground temperatures, in which conductivity or heat capacity is referred to without any realistic assessment of these parameters.

The varying degrees of suitability of soils for early germination, or early cropping is well known to farmers (see Taylor's paper and Chang 1968). That there are different times in any one locality, at which threshold temperatures at relevant depths for plant activity are reached is easily understood. Quantitative description or prediction may be difficult. The situation is often complex. If for example, a peat soil has a dry surface layer, temperature rise is rapid in that layer. Its thermal characteristics of low conductivity and low heat capacity resemble those of dry sand. Temperature change will be rapidly dampened with depth in the underlying saturated material with its high heat capacity. The difference in the energy exchange at the surface of dry, as compared with wet peat (Chapter 8) is also important.

A further clear example of the role of the thermal properties of the materials in the ground comes from highway practice. Sometimes a layer of synthetic insulation material is placed in a highway foundation to retard penetration of frost. Surface icing of such a highway may occur exclusively on stretches underlain by this insulating layer, because it retards the flow of heat in the autum upwards to the road surface. The road surface is therefore colder than adjacent stretches which are at temperatures above freezing, during night frosts.

Separating the effects of the nature of the ground surface (vegetation etc.) from those due to the thermal properties of the soil materials below the surface is not easy. The observations shown in Tables 10.1a and 10.1b relate to two sites significantly different only in soil conditions. The ground surface (the highway surface) and the above ground climate were closely similar. The figures illustrate the importance of soil thermal properties on the ground thermal regime.

10.6 Seasonal and other ephemeral freezing of the ground

Even in tropical regions, the surface of the ground may fall below freezing point, especially at high altitudes. Where night-time freezing is followed by daytime thawing, diurnal freeze-thaw is said to occur. In accordance with [10.3] and because of the effectively very low diffusivity due to latent heat (giving the soil a large apparent heat capacity – chapter 7.10), such freezing involves at most the upper few cm of the ground. But in regions with prolonged winter freezing a much greater depth will be involved.

The factors involved in penetration of frost into the ground are quite complicated and in spite of its importance for many engineering purposes no simple method of prediction has met with consistent success. More complex, potentially accurate methods suffer from a requirement for much information about the site, soil conditions, etc.

The main factors and their relative importance will be reviewed:

1. General climate: Cooling to produce freezing is often 'measured' by the product of negative (freezing) air temperatures and time, which gives the *frost index*. If an average freezing (air) temperature of -2 °C occurs for one day this gives $48 -$ °C h. Summing the corresponding figures for all days with freezing temperatures one obtains the freezing index for a winter. A value for example, for Ottawa,

Canada is 25 000 $-$°C h, for Oslo, Norway 12 500 $-$ °C h. There are several procedures for averaging and accounting for intervening periods of positive temperatures, giving slightly differing values. Further difficulties arise because the air temperature does not correspond exactly to ground surface temperature, and for example a thick layer of insulating vegetation or snow may on occasion prevent the occurrence of any frost index at the ground boundary.

In addition it is important to realize that winter freezing is superimposed on ground which has a quantity of stored summer heat at temperatures substantially above freezing (see Fig. 10.2, for example curve 5 and Table 10.1). In a continental type climate the hot summers result in frost penetrating to a smaller depth for a given frost index, compared to more maritime climates where there is less summer heating (Williams 1961). For comparable sites, frost penetration at Ottawa is only about half that at Oslo although the frost index is twice as great (Joynt and Williams 1973). For this, and other reasons the use of the frost index in relation to frost penetration requires considerable caution.

It appears that the maximum depth to which annual winter freezing commonly extends anywhere is about 4 m, although no doubt unusual circumstances might combine rarely to give a greater depth. Generally frost penetration occurs to increasingly greater depths as the mean annual temperature approaches 0 °C and as the frost index increases. Because of the symmetry of the annual temperature wave if the mean annual temperature and the frost index are considered together the importance of summer temperatures is (to some extent) taken into account at dispersed locations.

Thus if the frost index is used as a basis for prediction, it should be used in conjunction with knowledge of the mean annual temperature. However the frost index alone is a guide to the year by year depth variation at any one site, even though the year by year variations in snow cover may be more important.

2. Soil thermal properties: To consider this factor, we will again assume that we have in some way established the pattern of temperature change with time throughout the year, at the ground boundary.

The latent heat of fusion for the water present in the soil is the heat quantity to be extracted for freezing to occur (Chapter 7). This quantity for each increment of soil depth frozen, will be particularly large where water migrates to the freezing soil. But the heat capacity of the soil components cannot be ignored particularly where the ground temperatures are, only a metre or two below the surface, substantially above 0 °C due to summer heating as noted above. The heat represented by these temperatures depends on the heat capacity, and it must also leave the ground through the surface during the winter. In climates with warm summers, it may constitute most of the heat leaving the ground, even throughout a winter freezing period of several months duration.

The volumetric water content of a saturated sand is likely to be only about 30 per cent (of total volume) while that of clay may be 60 per cent. Thus the amount of latent heat to be extracted to freeze equal thicknesses of sands and clays can be in the ratio 1:2. Much of the water in clays freezes only if temperatures fall several degrees below 0 °C (Chapter 7), which reduces the ratio. However, if very dry soils, and soils of very high water content (e.g. peat) are compared, the ratio could easily be 1 : 3 or more. Of course, in nature these differences tend to produce large variations in frost penetration. Brown (1964) has suggested that in engineering practice, other conditions being equal the range of frost penetration associated with different soil characteristics is roughly 1:2. Tsytovich (1973) suggests a similar figure. The different heat capacities and quantities of latent heat are not the only factor involved even where soil type is the only variable. The conductivity of the frozen soil through which heat must flow upwards to produce further freezing is also important and often difficult to assess.

The spread of possible values for frost penetration in any one region is thus great indeed: special surface and soil conditions may combine to produce almost no frost penetration, while maximum values can easily be twice the mean value for a locality. This is why arbitrary acceptance of mean or 'typical' values can lead to engineering problems, when frost reaches to unexpected depths.

In an engineering context equations of the following type are often advocated:

$$Z_f = \beta \sqrt{\frac{2\lambda_f F}{L}} \qquad\qquad [10.6]$$

where Z_f = depth reached by freezing
λ_f = thermal conductivity, frozen soil
L = latent heat of freezing, per unit vol. soil
F = frost index
β = a coefficient (usual 0.5 to 1.0)

Although based on sound relationships involving effects of surface temperature change on homogeneous bodies of initially uniform temperature, such equations are open to strong objection in relation to frost penetration. The ground, of course, is not at an initially uniform temperature. In some situations, as noted, more than 80 per cent of the heat leaving through the ground surface has its origin in layers below the frozen layer. Equation [10.6] takes no account of the heat capacity but only of the latent heat of the soil. Although it permits some comparison of depths of frost penetration within areas of not too dissimilar climate, it does not appear to have a sound scientific basis for general application.

More sophisticated approaches to frost penetration, and the corresponding problem of summer thaw penetration in permafrost areas, have been made by several authors. If the frost index is to be used as the main parameter of winter climate, then some recognition of the role of summer temperatures and stored ground heat, as in Skaven-Haug (1972), is essential. Other methods involve determination of the heat flows through the ground surface, by examination of the surface energy and mass exchange processes (e.g. Scott 1964). The complexity of these methods, and the fact that the heat entering or leaving the ground is small in relation to other components of the energy balance, are difficulties. However, Smith and Tvede (1977) have successfully used commonly available climatological parameters for this purpose. Current research involves the relative importance of the individual parameters, and investigations of the extent of natural variations in soil thermal properties. Improved prediction of frost penetration would permit more economic design methods for highway construction and other purposes. Recent research has included consideration of heat and moisture transport in freezing soils as coupled phenomena (e.g. Guymon and Luthin 1974).

10.7 Water flows in the near-surface layers

Water in the form of rain, snow or vapour is a component of the above ground climate, and it follows logically that water and its movement constitute part of the sub-surface or soil climate. Flows of water, and of heat energy in soils are interrelated, and in a fully comprehensive analysis it is not possible to consider the one independently of the other.

The importance of heat storage and flow in the near-surface layers has been pointed out in this and earlier chapters. Water flow and storage relates to the question of availability of water for plants and is the subject of intense study for agricultural purposes. The near-surface layers which are not saturated are also important in hydrology. In questions of recharge of aquifers, the amount of water retained in or released from the near-surface layers is an important factor (Vachaud, Vauclin and Haverkamp 1975).

The subject can be approached on similar lines to that of heat exchange. During the course of a year the water content of the upper soil layer commonly undergoes a cyclic change as illustrated by Fig. 10.4(a), the ground being wetter in the winter than in the summer (in the

upper 2 m). If the soil in the profile is not uniform an irregular moisture content distribution with depth occurs. In such cases smoother curves may be obtained if instead of moisture content, a plot (Fig. 10.5a, compare Fig. 10.4a) is made of porewater pressure (or the suction, if the soil is not saturated). If conditions are of complete equilibrium, the relationship would be a straight line (Fig. 10.5). The suction and suction gradients are the key to understanding the moisture content distribution and movement within the unsaturated layers, just as the (positive) porewater pressures are the key to water movement below the water table. There is in this respect a close analogy with the thermal situation. Heat flow is given by the equation:

$$Q = \frac{\Delta T}{\Delta Z} \lambda$$

where λ = thermal conductivity
$\frac{\Delta T}{\Delta Z}$ = temperature gradient

The analogous equation for water movement q', cm s^{-1} has the form:

$$q' = k \frac{du}{dZ} \qquad [10.7]$$

where u = porewater pressure as cm water head
Z = distance
k = permeability (or hydraulic conductivity) cm s^{-1}

This equation is essentially [6.5], where q was in cm^{-3} s^{-1}. In [10.7] q' corresponds numerically to the volume of water which will pass through unit cross-sectional area in unit time; q' is the velocity of the volume flux (Rose 1966). The flow of water in g cm^{-2} s^{-1} would be given by $\rho q'$.

Considering the water movement in a *vertical direction only* we can write:

$$q' = k \frac{du}{dZ} \pm k \qquad [10.8]$$

If the soil is not saturated, u is the tensiometer potential. It is equal to the *in situ*

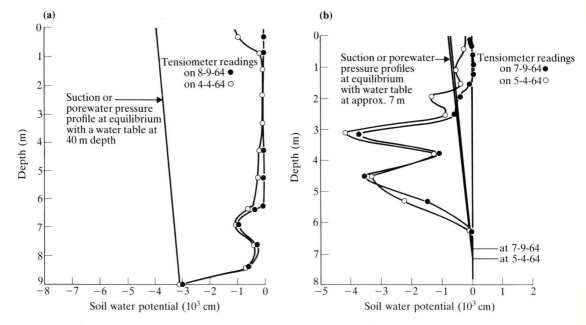

Fig. 10.5 (a) and **(b)** Suction profiles (matric potential profiles expressed as 10^3 cm watercolumn) for the soils in Figure 10.4 (a) and (b) respectively. (After Baver, Gardner and Gardner 1972)

suction, if the pore-air pressure is atmospheric. Clearly there will commonly be a steep suction gradient, if for example substantial evaporation is occurring through the soil surface. The last term has a positive sign if the flow is in a downward direction, negative if upwards. That term represents the effect of the positive or negative gravitational component potential.

A 'bulge' in the suction profile represents the passage of water which infiltrated through the surface during a period of rain (or perhaps irrigation), or alternatively a period of dryness (Fig. 10.5b). As in the case of a short-term temperature fluctuation the irregularity in the curve becomes less marked as the water, or the deficiency of water, with time proceeds to greater depth. The variations of moisture content and potential are however less predictable than those of temperature.

If the moisture content of a compressible, saturated soil (i.e. one without air in the pores) changes, the void ratio and thus the hydraulic conductivity change usually only slightly. The change in hydraulic conductivity (permeability) may be only some 10–25 per cent, which is

often not significant. But in unsaturated soils the hydraulic conductivity is highly dependent on the water content as noted in Section (6.12), and changes through orders of magnitude for the range of suction frequently occurring in near surface soils. Thus for partly saturated soils, equations [10.7] and [10.8] have limited meaning: any flow occurring will change the moisture content, which in accordance with the suction moisture content relationship must immediately change u. Note that the suction can be measured in cm water head, and it is convenient to use the symbol u also for the suction, in the present context. The value of u will, of course, be negative. Equation [10.8] can then be replaced by:

$$q' = k_\theta \frac{\partial u}{\partial \theta} \frac{\partial \theta}{\partial Z} \pm k_\theta \qquad [10.9]$$

(Cf. [6.8].

The term $\frac{\partial u}{\partial \theta}$ is the slope of the curve of the suction-moisture content relationship at moisture content θ. The product $k_\theta \frac{\partial u}{\partial \theta} = D$, the soil water diffusivity (p. 90), is clearly of fundamental importance where flow occurs in unsaturated soils. While D depends on water content, the phenomenon of hysteresis in suction-moisture content curves means that $\dfrac{\partial u}{\partial \theta}$, and thus D, has more than one possible value for a particular water content.

This circumstance is considered by Staple (1969) in a study of moisture content distribution in initially dry soils. In studies of change of moisture content a flow equation is used (the suction effect is better represented by the symbol ψ, rather than u):

$$\frac{\partial \theta}{\partial t} = \frac{\partial}{\partial Z}(k_\theta \frac{\partial \psi}{\partial Z}) \pm \frac{\partial k_\theta}{\partial Z}\theta \qquad [10.10]$$

where $\psi =$ potential (determined from suction moisture content test)

$t =$ time

Staple determined the relationships of ψ and k to θ, and used a finite increment method of computation to predict moisture content distributions following infiltrations of different amounts of water applied to the surface of soil columns. The computed values agreed well with observations (Fig. 10.6).

10.8 Water storage and the water balance in the near-surface layers

In other respects too, the question of water exchange must be approached differently to that of heat exchange. On average the annual intake of heat through the ground surface is balanced by an equal output (except in the case of microclimatic or climatic change). The geothermal gradient provides, in this context, an insignificant flow of heat. In the case of the water balance, in humid climates water entering the ground surface and infiltrating downwards exceeds that lost by evaporation on an annual basis. In dry climates the reverse is the case.

(a)

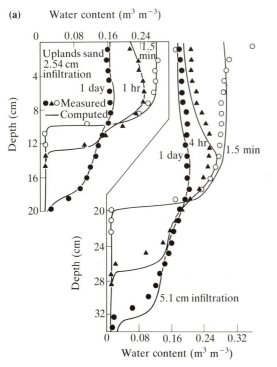

(b)

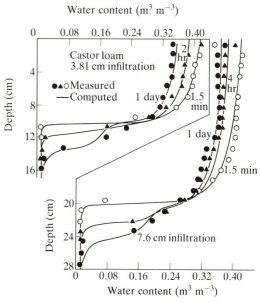

Fig. 10.6 Soil moisture redistribution profiles for (above) Uplands sand and (below) Castor loam. A layer of water, of depth shown, was applied to the tops of soil columns. Moisture contents were then determined during the course of infiltration, after times as indicated. (After Staple 1969)

It is the ground water (the reservoir at depth) which receives the excess intake, or supplies the excess output. Thus considered over the year we find either a net upward or a net downward flow of water through the near surface layers. Which of the two circumstances applies, and the magnitude of the net flow, has importance for example in relation to soil and rock weathering. The quantity of water entering or leaving the reservoir of ground water is of course of prime concern to ground water geologists and irrigation hydrologists.

In spite of the fact of a net upward or downward passage of water through the near-surface layers above the water table, the average quantity of water held in those layers remains essentially unchanged on a year to year basis. The average water content throughout a year can be considered as a parameter similar to mean annual ground temperature. It continues essentially the same unless there is a change of climatic conditions, or several years

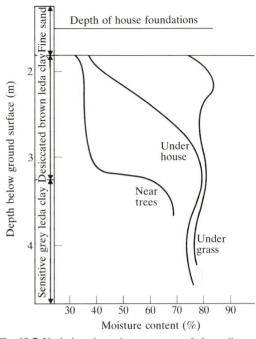

Fig. 10.7 Variations in moisture content of clay soil at closely adjacent sites during summer. (After Crawford 1968)

Plate 10.1 Late Victorian house in Ottawa, showing settlement due to soil volume changes. The weight of the house on its foundations being insufficient to cause much consolidation of the clay soil, the shrinkage effects are probably due to the lowering of ground water tables and drying of the soil, particularly in association with the provision of drainage. (Photo reproduced by courtesy of National Research Council, Canada.)

of unusual weather, or a change in the nature of the ground surface. The effects of different surface covers are illustrated in Fig. 10.7. Just as changes in the soil thermal regime may have far reaching consequences for example in areas of permafrost, changes in the moisture regime whether naturally or artifically induced may have great practical importance. In addition to agricultural applications of drainage or irrigation, changes associated with urban development (with improved drainage for example) can lead to problems of soil shrinkage or alternatively of swelling, with ensuing damage to foundations (Pl. 10.1) (Bozozuk 1962). Much work is currently being carried out on infiltration and the distribution of moisture at different depths and times in the unsaturated layers, and a detailed older review is that of Luthin (1959).

It appears that precise prediction of moisture movements will always be difficult in view of the circumstances described and the variability of natural soils. The relationship of soil moisture behaviour to thermal conditions, whether surface evaporation, vapour transfer within the soil, or freezing, further complicates such analysis.

11

Periglacial forms and processes

11.1 The periglacial regions and frost climates

About 2 per cent of the northern hemisphere land area is glaciated, or perennially covered with ice and snow. A much larger area is described as periglacial (peri = around, hence 'nearly-glacial'). The presence of glaciers follows from a combination of sufficient precipitation and low temperature; many periglacial areas experience lower temperatures than some glaciated areas. The loosely-defined term 'periglacial' relates to climates which because of freezing and thawing give distinct surface and subsurface forms and processes, and a distinct, often sparse vegetation cover. Geomorphologically, thawing during the summer may be as significant as the coldness of the winters.

The etymological association with glaciation is unfortunate; the distingushing periglacial features often occur hundreds of kilometres from glaciers, while in other areas they occur barely further than a few hundred metres from glacier margins. In addition, some of those areas where the effects of freezing and thawing have come to be of great technological and economic significance (some of the most populous parts of Canada and Russia for example), are not generally regarded as periglacial at all.

The traditional view of the periglacial region relates on the one hand, to mountains, or 'alpine' areas, where high altitude results in low temperatures, even in middle or lower latitudes. On the other are the 'high arctic' (high latitude) areas (and the 'high antarctic' areas – although both the knowledge and extent of the land areas is there more limited). Both environments are characterized by lack of trees, tundra conditions, and by the presence of particular microrelief forms, or ground surface formations, known generally as patterned ground. There are also various, characteristic, relatively shallow, movements of soil downslope known as solifluction. The two environments, high latitude and high altitude, differ in that permafrost (perennially frozen ground) occurs in the high latitude areas rather generally, while it is commonly absent in the middle and low latitude mountainous periglacial areas. The latter include much of Scandinavia, and the mountain chains of middle North America, Switzerland, Austria, and certain African regions.

The boreal forests and flat peatlands covering vast areas in northern Canada and central northern Europe appear totally different to the tundra. Yet, at least where underlain by permafrost, these terrains also present important characteristics associated with freezing and thawing. It is particularly because of economic development in the northlands, that these, in many ways less attractive, regions are only now becoming part of the periglacial regions as described in textbooks (e.g. French 1976).

Clearly the periglacial regions are not to be identified with a single special kind of terrain or of topography. Major relief forms, valleys, mountains and plains, generally have their counterparts elsewhere in the world where freezing does not occur significantly. Generally periglacial conditions have of course, occurred only over a geomorphologically-speaking short (i.e. post-glacial) timespan of thousands or tens of thousands of years. The preceding glacial period was a hundred or so times longer, and even that did not erase much of the topographic relief predating it.

Periglacial conditions and the effects of freezing temperatures in the ground are currently of great geotechnical interest, because of the problems of building highways, pipelines, and other structures (Crawford and Johnston 1971). In many respects these geotechnical concerns are common to all the periglacial terrains, but the occurrence of permafrost in addition to seasonal freezing and thawing raises special problems.

There is a frequent misconception that the smaller-scale surface features of periglacial regions, patterned ground and solifluction, are necessarily associated with the occurrence of permafrost. Deep annual freezing is sufficient to produce many of these features, which are discussed in this chapter. The term periglacial is by no means synonymous with permafrost. Permafrost has been estimated to underlie (although not entirely continuously) some 25 per cent of the earth's land area. The periglacial regions, loosely defined though these are, cover probably half as much again. Identifying 'periglacial' with the effects of freezing and thawing, as in this chapter, is not intended as a validification of 'periglacial' as a term of regional description. It serves however to emphasize that freezing and thawing are commonly a major element in the behaviour of the earth's surface.

11.2 Permafrost

Permafrost is simply ground which remains frozen for more than one year. One might attempt to narrow the definition – for example to exclude frozen ground that happens to survive one particularly cold summer – but difficulties arise. Permafrost is often far from 'permanent'. Its occurrence, however, is to be understood in a different way from seasonal freezing.

Where the mean ground temperature is less than 0 °C, then below the depth of annual temperature variation, the ground must be frozen perennially. Permafrost will also occur at shallower depths, where the annual variation does not include positive (above 0 °C) temperatures. Contrary to loosely-expressed remarks in many papers, the *existence* of permafrost is not a matter of a 'negative heat balance of the ground'. As noted in Chapter 10, the mean ground temperature varies with location. Whether or not the annual heat balance (using the term in the sense discussed on p. 34) is positive (gradual warming) or negative (gradual cooling) is a separate question, which however is very important for increase or decrease (*aggradation* or *degradation*) of the permafrost at a particular place.

In the far North, permafrost is more than

600 metres thick. As the thickness decreases southwards, the *active layer* above the permafrost, which freezes and thaws each year, tends to increase (Fig. 11.1). It varies from a few cm or less under certain surface conditions in the far north to several metres (also depending on surface conditions) at the southern limits of permafrost. The thickness and lateral extent of permafrost towards the southern limits of its occurrence varies greatly within short distances. There is for example no permafrost below the Mackenzie river in Canada, even though in its northernmost stretches, the permafrost is at least a hundred metres thick on either side of the river. Further south, the lateral extent of land underlain by permafrost decreases, until ultimately there are only a few, widely scattered, bodies (Brown 1967). The term 'discontinuous permafrost' distinguishes such regions from those of 'continuous permafrost' even though it is not possible to demarcate unequivocally the regions. There may be a single hole or 'slit' through the permafrost, as occurs below the Mackenzie river, isolated by tens of kilometres of otherwise continuous and thick permafrost.

These distributions are easily understood in a general way. The temperature level of the ground, more precisely the mean ground temperature, depends upon climatic conditions and energy exchanges between the ground and atmosphere. We have seen (section 10.3) that there is a rough correlation between the mean annual air temperature and the mean annual ground temperature, but also that the mean

annual air temperature is usually a few degrees, and sometimes ten degrees, lower than the mean annual ground temperature. Even more important, the difference itself varies by a degree or more over small distances (Fig. 11.2). Herein lies the clue to discontinuous lateral distribution of permafrost.

Consider a locality, where a few measurements reveal mean ground temperatures near but slightly above 0 °C. Suppose that there is normally a snow cover, but that at an exposed site snow is consistently blown away. During the winter one would expect to record somewhat lower temperatures (in the near surface layers) there because not only is the site windy and 'cold', but the insulating blanket of snow is absent. If during the winter the ground is colder, the mean annual temperature will also probably be somewhat lower there, and, if it is less than 0 °C, permafrost occurs. In regions of mean ground temperature near to 0 °C, permafrost is often found under such exposed situations while absent elsewhere in the vicinity.

There are many variations of surface conditions affecting mean annual ground temperatures in any one locality, as discussed in Chapter 8. For example, during the summer, particularly wet ground, or particularly abundant vegetation, are often associated with rather lower summer temperatures. A relatively

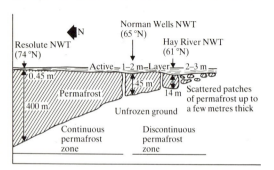

Fig. 11.1 North-South vertical profile of permafrost in Canada showing decrease in thickness southward, and in relation to continuous and discontinuous permafrost zones. (After Williams 1979)

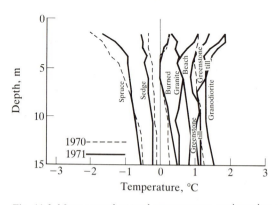

Fig. 11.2 Mean annual ground temperatures, various sites at Yellowknife, NWT, Canada. The sites are broadly characterised according to surface cover or bedrock. Variations at particular sites, especially at shallow depth, are presumably related to the changing thermal properties of the active layer during the course of the year (see p. 151). (After Brown 1973)

large proportion of the net radiation is being utilized in evaporation, reducing that energy available for warming the ground. This in turn lowers mean annual ground temperatures and favours permafrost (unless perhaps, the effect is offset by especially thick snow accumulations and warmer winter ground temperatures).

Since the exchange of heat at the ground surface involves so many processes which are controlled by the nature of the ground surface, relief, exposure, vegetation etc., it is difficult to assess quantitatively the various component fluxes (Chapter 8) and whether they in combination produce freezing mean ground temperatures. This is particularly so where ground temperatures in any case are close to 0 °C. Where the mean annual ground temperatures are, in general, several degrees below 0 °C only quite extreme surface conditions will result in an absence of permafrost. Large water bodies, such as the Mackenzie river, are an example. If the water does not freeze to the bottom in winter, the ground surface (that is, the river or lake bed) is never frozen. If the water body is wide enough the permafrost on opposite sides will not meet at any depth under the water body.

In localities of discontinuous permafrost, an observant person, particularly if he has a basic understanding of the phenomenon, can with experience often pick out areas where permafrost does or does not occur merely by observing terrain characteristics. Unfortunately, experience gained in one locality cannot always be transferred to another, such is the varied nature of terrain and climatic conditions, and the many variables in the energy exchange.

Few detailed studies relating surface energy exchange conditions and mean ground temperatures have been made (but see Smith 1975) and consequently the relationship of discontinuous permafrost to surface conditions is for the most part known only in a qualitative fashion.

The position of the lower surface of permafrost is commonly determined by the geothermal gradient as shown in Fig. 11.3. But this gradient may itself be disturbed by climatically or microclimatically-induced

surface temperature changes in the past (p. 152 and Judge 1973), so that linear extrapolation downwards to the 0 °C level may be uncertain. Under 'steady-state' conditions the depth of the lower surface of the permafrost will, of course, reflect differences of mean ground temperature due to different surface conditions (Fig. 11.3).

11.3 Permafrost aggradation and degradation

Changes over time at the ground surface, at a particular site, must affect the energy exchange. The sub-surface heat fluxes are likely to be modified. For example, bared ground will absorb more heat during the summer, and during the winter (assuming the snow cover remains) there is unlikely to be a counter-balancing increased heat loss. The heat budget of the ground determined year by year will show temporarily a positive balance – the

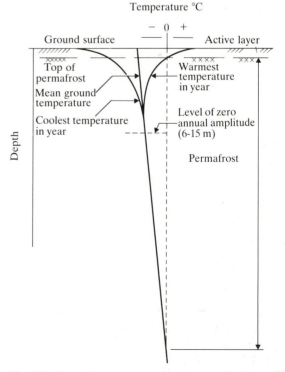

Fig. 11.3 Ground temperature regime in permafrost. (After Williams 1979)

temperature of the ground (the mean annual ground temperature) will rise. There will be degradation of permafrost. This may merely involve a slight deepening of the active layer, but if the permafrost as a whole has temperatures only just below thawing point degradation may continue over many years, until perhaps all is thawed. In regions of discontinuous permafrost, recurring changes in ground surface conditions (e.g. forest fires, the natural succession of vegetation, disturbance of surface drainage) result in changes over some years in lateral distribution of permafrost. If conditions of general stability (lack of change) persist at the ground surface, ultimately a more or less stable configuration of the permafrost occurs. This implies, of course, that there is neither a continuing negative, or positive, balance of heat year by year leaving or entering the ground.

Conditions at the ground surface are not, in fact, ever completely stable, and in addition climatic change must be regarded as 'normal'. Consequently stability of discontinuous permafrost is at best relative. The important principle remains that ground energy balances, whether positive or negative are associated with aggradation or degradation, while the stable existence of permafrost is associated with energy balances on a yearly basis averaging out to approximately zero, that is with quasi-equilibrium climatic and ground surface conditions. Of course, in regions, where mean annual temperatures are well below 0 °C and permafrost is continuous, climatic or microclimatic changes will affect only the upper layer of permafrost and the depth of the active layer.

Just as seasonal temperature fluctuations are propagated slowly into the ground to a certain depth so are climate or ground microclimatic changes of longer duration felt progressively at greater depths. Suppose that for a number of years warmer, 'non-permafrost' climatic conditions prevail. The temperature effect penetrates the ground at a rate more or less proportional to the square root of time from onset. The warmer period is then followed, we will suppose, by a cooler one, which gives a

gradual cooling spreading from the surface. As a result at some depth the ground will be warmer than either above or below. This warm layer constitutes a bulge in the temperature gradient (the geothermal gradient, as this was defined p. 152). If temperatures are in any case near 0 °C this may result in a layer which remains unfrozen for a long period. Such a layer of unfrozen ground sandwiched in the permafrost is a *talik* (Fig. 11.1). It may also lie on top of the permafrost immediately below the active layer (the deepest point reached by seasonal freezing). Sometimes the talik is thicker than the permafrost.

The 'inverse' of the talik is the isolated layer of frozen material. If a permafrost layer occurs only at some considerable depth, below the depth reached by annual temperature changes, then it is probably only a matter of time before it thaws bringing that body of soil once again more in line with a uniform temperature gradient. Such permafrost is called '*relict*'. It is not in thermal equilibrium with its surroundings, and its persistence can be likened to the relatively long life of ice cubes in a drink. The disappearance of relict permafrost may be interrupted by the onset of colder conditions again. The large amount of latent heat that has to be supplied for thawing to occur, is normally transferred only at a rate determined by the temperature gradient and the soil conductivity. Sometimes the process is accelerated by water infiltrating into the ground. Isolated small bodies of permafrost, perhaps only a few cm in thickness may occur just below the active layer. These are due to recent climatic or microclimatic changes, perhaps merely an unusually cold year.

11.4 Structure of permafrost

Visible segregations of more or less pure ice are abundant in permafrost. The ice segregation or frost heave processes discussed in Chapter 7, are important, and certain special conditions apply. When permafrost forms at depth freezing is slower than is the case for the active layer during the winter. Slower freezing is conducive to large segregations, in that, because of the length of time available, the low permeability of fine-grained, clay soils ceases to be a limiting factor. At depth the soils are generally saturated, a circumstance aiding moisture transfer. These effects are counteracted however by the greater effective stress (total stress less water pressure) occurring as depth increases. Indeed, the increasing effective stress provides a limiting factor for segregation as described in section 7.9. According to one study (Williams 1968) silty soils show little ice segregation at depths greater than 20 m. Ice segregation occurs at greater depths in clay-rich soils. This is partly due to consolidation of clay between ice layers, with a strictly local transfer of water to the latter. The process can continue at temperatures significantly below 0 °C, when correspondingly high suctions (Fig. 7.6 and p. 97) are developed in the unfrozen water of the frozen clay. In addition some migration of water from adjacent unfrozen ground, through the frost line, is possible because of the permeability of frozen soils. The effects of the gradients of suction (or potential) within the frozen zone extend into adjacent unfrozen material. The suctions could be large enough that frost heave (increase of total moisture content by ice segregation) would occur in spite of the large overburden pressures at depths of tens or even hundreds of metres. As yet there is insufficient evidence for firm conclusions.

Ice bodies have various origins (reviewed by Mackay and Black 1973). These include glacier ice or snow buried during deposition of sedimentary material or till, ice in veins or cracks perhaps originating from surface water, and ice in ice wedges as discussed in section 11.6. Permafrost in coarse-grained material will be free of segregated (frost-heave) ice, but may contain ice bodies of these other kinds. Although the amount of 'excess' ice, i.e. in bodies larger than pore size, varies greatly in all soils, it is often particularly abundant in low-lying, wet, areas.

The lack of permanency of permafrost in relation to thermal effects has been stressed

and will be considered further subsequently, in relation to particular ground surface features. It is also unstable in a somewhat different sense and with respect to its composition. Permafrost is a material with one component at its melting point, in that ice and water coexist in frozen soil at least to temperatures several degrees below 0 °C. The proportions of ice and water change with temperature and pressure. Thawing of some ice, migration of water, and refreezing occur in association with temperature or pressure gradients. Additionally recrystallization (without intermediate thawing) is an important aspect of the rheological properties (see p. 179). Harlan (1974) has pointed out that the geothermal gradient represents a gradient of free energy or potential in an upward direction. The movements of water, and the redistribution of ice, to be expected on this account are to some extent offset by the gravitational potential gradient but not completely. There is a gradient of total potential upwards. The implication is that, on a geological time scale, or perhaps one of only tens or hundreds of years ice will concentrate in the upper parts of a permafrost layer.

Erosion and the development of slopes produces new temperature and stress distributions which apart from mass movements, landslides, and settlements, associated with total thaw, may also lead to long term redistribution of moisture content within the permafrost. The special geotechnical problems of permafrost arise because of the melting point relations in the broadest sense. No aircraft designer, for example, would use a material whose operating temperature would be within a few per cent of the melting point on the absolute temperature scale. Yet this is what the geotechnical engineer in cold regions is consistently expected to do in his construction activities.

11.5 Pingoes and palsas

There are three groups of features whose formation necessarily involve permafrost, and which therefore are diagnostic of permafrost conditions. The most widespread are ice wedge polygons, which are considered under 'patterned ground' (next section). Quite the most individually conspicuous features whose occurrence is unquestionably due to freezing processes are *pingoes*. Their complex mode of formation and association with permafrost were earlier the subject of much speculation. *Palsas* by comparison appear to be essentially due to frost heave in the formation of small bodies of permafrost.

Pingoes are rounded or somewhat conical hills often tens of metres high, and filled by a core of ice. Pingoes are of two kinds – the 'Mackenzie valley' type, which are found in low-lying wet areas; and the 'Greenland' type which are usually at the foot of slopes. Both types develop mainly by gradual growth of the ice mass producing the hill. Mackay (1973) who has carried out comprehensive theoretical and field investigations reports a growth rate up to 1.5 m yr^{-1}. Changes in climate or perhaps the changed conditions resulting from the gradual elevation of the pingo top above surrounding terrain, result ultimately in decay. The top of large pingoes is commonly gullied. A number of crater-like ponds, for example, in northern Sweden and Great Britain are thought to be the remains of totally thawed pingoes.

The 'Mackenzie valley' type are initiated by the gradual freezing of the body of unfrozen ground underlying a receding lake in a permafrost region. As the lake shoals over or drains, the mean ground temperature falls and winter freezing of the bed occurs. Gradually permafrost forms, spreading downwards from the surface, and also encroaching laterally (Fig. 11.4). Depending on whether the sediments are coarse-grained or fine-grained two distinct situations occur in relation to the soil water. Although the situations are opposite in their immediate effect, it is likely that both have a role in the growth of the pingo.

As Mackay (1973) points out, where the sediments are somewhat coarse-grained and particularly where overburden is substantial, ice formation and the associated volume increase pushes water away from the front line. The pore water pressure will tend to rise, and the effect will spread in the unfrozen material. Uplift of the strong, newly-frozen layer on the lake bed as a result of these pore pressures alone seems unlikely because of its strength and thickness. However, if finer material also occurs this is liable to frost heave, and the elevated water pressures will make this more pronounced, particularly where the overburden pressure is not too large. For such finer soil materials the suction generated (see p. 104 and [7.10]) when added to the already elevated pore pressure, results in flow towards the frost line

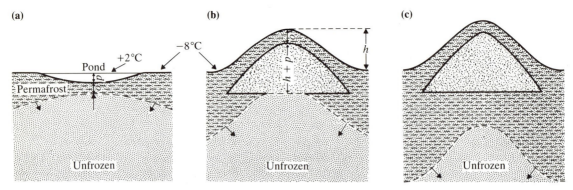

Fig. 11.4 Growth of pingo (a) Growth commences under a residual pond of depth *p* and with a layer of permafrost thickness *o* which becomes overburden above the ice core. In (b) core has grown to a thickness *h* + p, this being the height of the pingo above the original pond level. Following a period with a stationary frost line at the base of the ice core, further uplift may continue by frost heave and ice segregation as the frost line penetrates deeper (c). Continued growth may involve freezing of pore water in a confined system (After Mackay 1973)

where the ice core develops. The enlargement of the ice core then occurs with a pressure in the ice sufficient to lift the frozen surface layer and ultimately perhaps many metres of ice and frozen soil (Williams 1967, p. 109). Thus the growth of pingoes is apparently a combination of the effects of extrusion of water on freezing of coarse-pored material in a more or less closed system, and the transference of such water to the ice core by the frost heave (ice segregation) phenomenon this being particularly relevant with finer-grained material.

In the Mackenzie Valley there is often substantial interbedding of coarse and fine material, providing the conditions for such a sequence of events. Mackay's observations show the growth of pingoes to be primarily occurring below the centre of the ice core. A part of pingo growth is also ascribable to frost heave with formation of lenses or layers interspersed with soil.

The 'Greenland' type of pingo (Müller 1959) involves artesian pressures developed as a result of ground water trapped between the permafrost and a frozen surface layer. At the foot of the slopes the high water pressures are sufficient to lift the frozen surface layer, perhaps aided by a higher pressure in the ice due to the frost heave or ice-segregation process, and pingoes with ice-rich cores form.

Palsas (from the Swedish pals, pl. palsar) are relatively small mound or ridge-shaped features occurring especially in regions of very discontinuous permafrost (Lundqvist 1969). Often only a metre high, but sometimes 7 m or more, they occur most commonly in wet situations with a scrub or moor vegetation and a humus layer, 10–50 cm thick. The palsa in fact is an isolated permafrost body which has undergone considerable frost heave – the permafrost core is ice rich, and perhaps an ice mass with only thin discontinuous soil layers. The raised surface explains the permafrost condition, while the mean ground temperature is elsewhere above freezing. Once the localized frost heave is initiated, the winter snow cover is probably thinned by wind over the elevated

piece of ground, so that soil temperatures are lower. During the summer the elevated vegetation and humus is somewhat dryer than elsewhere, and the decreased thermal conductivity on that account means that summer heating of the ground is also less. A slow aggradation of the small permafrost body occurs with further ice accumulation until a size is reached at which extension of the permafrost core ceases. Thaw and collapse may follow as a result of microclimatic changes on the palsa surface itself. Although soils of a very peaty nature are usually involved, sufficient finer-grained mineral soil must be present for frost heave to occur. Palsas are an indicator of discontinuous permafrost conditions; they illustrate how surface conditions provoke aggradation or degradation of permafrost bodies.

11.6 Patterned ground

In northern Canada, and the Soviet Union, enormous areas show from the air a roughly polygonal patterning. Each 'polygon' is perhaps 10 m in width and delineated by ditch-like features with raised edges. The vegetation cover is often more or less continuous. This terrain represents the most widespread form of patterned ground – the term refers to a variety of repetitive surface and microrelief forms which are characteristic of periglacial, and particularly tundra conditions.

In the case of the polygonal features described the subsurface conditions are as striking and curious as the surface arrangements of ditches and ridges. Below the depth of the active layer in the ditches there are vertical wedges of ice, a metre or more across at the upper surface and extending several metres downwards (Fig. 11.5). This abundant form of ground ice is of major importance geotechnically insofar as the wedge may melt, with subsequent surface settlement. It also provides the specific name – *ice wedge polygons* – for this form of patterned ground.

Lachenbruch (1962, 1963) in a comprehensive study concludes (as do earlier

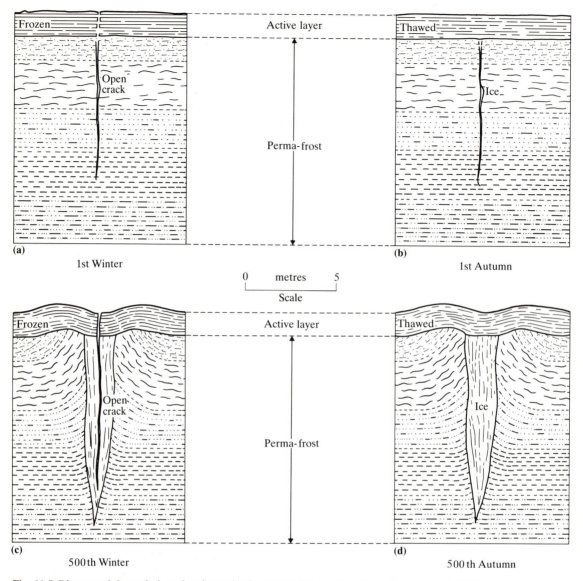

Fig. 11.5 Diagram of the evolution of an ice wedge by contraction cracking. (After Lachenbruch 1963)

authors) that ice wedges originate in vertical cracks associated with contraction of the ground during winter cooling. Ice has a coefficient of thermal contraction some five times that of soil mineral components, such that the greater the ice content of the soil the more likely it is to crack in this way. Because pure ice has a lower tensile strength than frozen soil, cracks form each winter within the ice wedges. During the summer period, the frozen ground goes into a state of compression (associated with thermal expansion and the increasing volume of ice represented by the wedges). The frozen soil and the thawed soil above tends to be pushed up to give the elevated 'banks'.

Lachenbruch points out that a viscous strain occurs during the winter contraction, so that rapid cooling is more likely to produce cracking. Slower cooling allows a slow strain, or 'give', of the ground (tending to give stress relaxation) such that the tensile stresses are less likely to reach a value sufficient for cracking to occur.

The magnitude of the temperature drop from the mean, and the rate of cooling necessary are not well-defined; in any case they will vary depending on the mechanical properties of the frozen soil in question. The temperature must be low enough that the expansion associated with transformation of water to ice, is exceeded by thermal contraction. For fine-grained soils this will be several degrees below zero. But the depth to which wedges reach, and the frequent presence of a surface cover reducing winter cooling (snow, and thick vegetation), means that a substantially greater drop in air temperature is required.

Initiation of a crack often makes a loud report. These have also been heard and cracks subsequently observed in regions without permafrost, during periods of unusual cold. Seismological instruments record the cracking. When a crack forms, stress is relieved in a direction normal to the crack but not parallel to it. Subsequent cracks are therefore more likely at right angles, or at least at some substantial angle to the original one.

The ice wedge 'polygons' are rarely perfectly symmetrical. Frequently a 'rectangular' pattern occurs. The different arrangements of the cracks are due to lateral variations in tensile strength and rheological properties of the frozen ground, and also probably to microclimatic differences, such that slope, moisture content gradients, vegetation changes and other differences are important. Fine-grained soils are probably more susceptible to ice wedge formation because of the additional tensile stresses in the soil between segregated ice layers, in association with the suction forces developed by freezing.

Field observations have shown annual rates of increase of thickness of wedges up to several mm. The sources of the moisture which penetrates the crack to enlarge the ice wedge are somewhat uncertain. These, and other points are reviewed in detail in Washburn (1979). Assuming the crack extends to the surface, snow and vapour can enter. If some part of the active layer is unfrozen then water may trickle in.

In arid or semi-arid cold regions, as those of Antarctica, cracks occur without becoming filled with ice. Instead soil particles fall into the crack giving rise to wedges of infilled soil, often distinct in grain size and structure from the *in situ* material.

In regions which no longer have a permafrost climate, 'fossil' wedges of infilled soil are found. These infillings may be material originating in the active layer, and which replaced ice wedges slowly melting from the top downwards in association with a warmer climate. More rarely their origin may be that just described for arid or semi-arid conditions.

The top of an ice wedge usually marks the bottom of the active layer. Sometimes ice wedges occur only at greater depths, this being due, for example, to surface deposition of sediment after the wedge has formed. Such buried ice wedges are evidence that, contrary to section 11.4, ground ice may on occasion persist relatively unaltered over hundreds or thousands of years.

A much less widespread form of patterned ground although one which has traditionally excited the curiosity of travellers, is *stone polygons*. Sometimes their regularity of form, with the arrangement of stones and boulders in regular polygonal patterns (each stone-bordered polygon is often several metres in diameter), seemed inevitably the work of man. Other patterned ground involves a distribution of vegetation, with 'holes' in a network, or perhaps discrete circles, of bare soil. Sometimes the central, bare soil patches have a flat dome shape. These, where ground conditions are uniform, cover larger areas. Surface hummocks are a further, widespread, if less striking form. Very large areas of organic soils are uniformly covered with hummocks, 75 cm or more high and of similar width, so closely spaced as to make walking difficult. The many different types of patterned ground were reviewed by

Washburn, (1956, 1979). In spite of a voluminous literature on the subject, there are few detailed studies of particular types, and their mode of origin is understood only in a general way. It is appropriate to review generally a number of processes which are involved in patterned ground formation.

Sorting, or the separation of stones from finer soil, is common to much patterned ground, and is the basis of Washburn's subdivision into 'sorted' and 'non-sorted' forms. It is a common observation in regions of substantial winter frost penetration that stones and boulders appear at the surface of the ground every spring. During frost penetration the higher thermal conductivity, and the much lower heat capacity of boulders (because they do not contain water), gives a high diffusivity such that freezing commences below a boulder earlier than at the same depth in the surrounding soil. Consequently as frost heave commences under the boulder there is a tendency for it to be pushed upward. The process is illustrated in Fig. 11.6a. Beskow, and others, believe the freezing soil pulls the boulder as well. During the spring thaw the ground immediately below the boulder thaws first. There is usually a layer of ice adjacent to the boulder on its lower side, and when this thaws a cavity is left (the boulder still being supported laterally). Gravel and other material falls into this cavity so that the boulder cannot drop down into exactly its original position. An alternative mechanism involves stones or objects with a long axis, which comes to be more or less vertical. Frozen soil 'grasps' the upper part of the stone, and subsequent heave results in the stone being pulled upward, relative to lower lying unfrozen soil (Fig. 11.6b). For the stone to remain elevated at thaw, it appears again that soil material must move into a space immediately below the stone. Washburn (1979) reviews many theories in detail.

Large stone polygons probably involve these processes, but with cracking of the ground, as in ice wedge polygons, probably providing for the positioning of the exposed boulders.

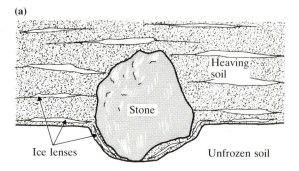

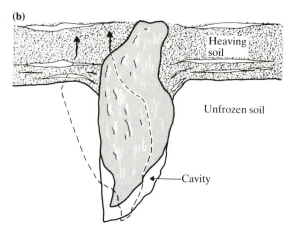

Fig. 11.6 (a) Displacement of stone by growth of ice lens beneath.
(b) Raising of stone to vertical position by frost heave

Assuming such cracking to occur, then there may be movement preferentially towards the cracks because frost heave in that direction is not limited by overburden pressure.

Internal erosion, or washing out of fine material occurs in the formation of certain patterned ground, but it seems unlikely to be an essential factor in the formation of stone polygons. Although the differential displacement of stones towards vertical cracks may intuitively seem less than completely plausible, there is no strong evidence of any other process having an equivalent role.

A different kind of sorting involves soil particles (Corte 1966, Fredén 1965, Rowell and Dillon 1972) of clay size and up to 1 mm particle diameter. Particles tend to be pushed ahead of an advancing ice-water interface, the process being related to that of ice segregation.

Because water close to particle surfaces has a lower freezing point and a corresponding suction, water tends to be drawn into the adsorbed layer between the ice and a particle. In this way the layer is maintained while further freezing occurs. A growing ice lens is separated from all adjacent particles by such layers. When ice advances instead into the pores, particles which are significantly smaller than most, and smaller than the pores, can be pushed ahead. They remain separated from the ice by the adsorbed water layer.

The role of the process in the ground is conjectural. Displacement of fine material through pores between coarser material may be important in the sorting often seen in periglacial soil profiles and involving irregular layers of grain size composition finer than adjacent material. Accumulations of fine material occur as well in a number of forms of patterned ground.

There are a variety of small, shallow forms of patterned ground, known as mud circles, non-sorted circles, stony earth circles, and by other names. These are often less than 1 m in diameter. One form is common on exposed lichen-heath areas with the vegetation being interrupted by stony patches, only a metre of two apart. Rather detailed investigations (Williams 1959) showed that wind erosion removes both snow and loosened surface materials to initiate bare spots. The frost then penetrates more rapidly, with a bowl-shaped frost line, below these spots. As heave and stone uplift occurs more or less perpendicularly to the frost line, there is a gradual concentration of stones at the bared surface. Ultimately there is an almost complete loss of vegetation from the sharply-defined stony patch. A single such circle persists for some tens of years, but will eventually be recolonized with vegetation.

Distribution of vegetation, and uneven frost penetration associated with different insulating effects of root or above ground growth, are important in hummock formation. Differential frost heave (of which stone uplift is only one form) results in unequal displacements of the soil and is presumably the dominant process. The initial formation of the hummock tends to increase differential heave.

In addition to the processes discussed various others have been invoked such as cracking due to drying (a likely explanation for certain 'micro' polygons), differential weathering leading to accumulation of fine material, and so-called 'cryostatic' pressures generated in more or less closed pockets or bodies of soil by encroaching frozen soil. Quite high pore-water pressures of the latter kind are sometimes observed between a penetrating layer of winter freezing and the top of the permafrost. Suffice it to say that most patterned ground forms are the product of a complex interaction of processes and situations. This is why there are so few complete descriptions of the formative processes. The processes are slow in effect so that the growth of the features can generally only be measured over periods of years. By contrast, distribution of the different forms is fairly well recorded and can serve as a guide to the causative environmental factors.

11.7 Solifluction

A feature of slopes which is characteristic of periglacial conditions is lobate or terrace-like surface forms developed in the upper metre or so of the soil by slow movement downslope (Price 1973). Viewed from the air the impression is of some viscous material flowing in lapping waves over the surface. On the ground the wave fronts appear as bulging ridges or banks. The banks often contain an accumulation of boulders, and may be a few cm or up to several metres high. Most commonly however they are less than 1 m in height, Fig. 11.7.

Downslope movements of up to several cm yr^{-1} are often recorded at the soil surface. Active movement is evidenced by inverted plants at the foot of the banks. The rate of movement decreases with depth, although as shown in the example (Fig. 11.8), the

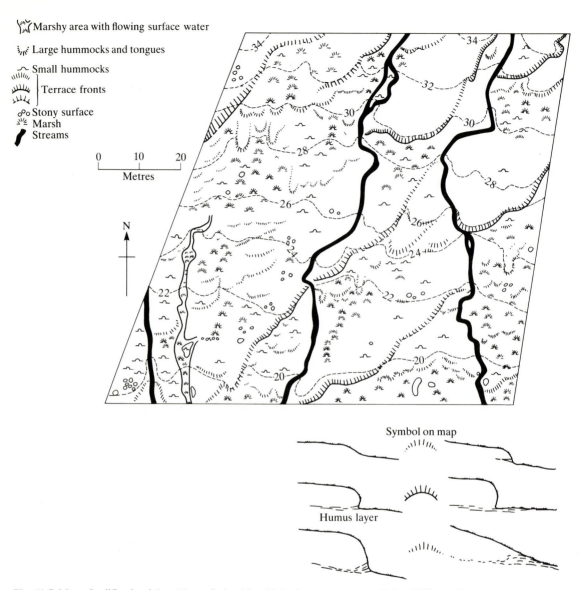

Marsh area with flowing surface water

Large hummocks and tongues

Small hummocks

Terrace fronts

Stony surface
Marsh
Streams

0   10   20
Metres

N

Symbol on map

Humus layer

Fig. 11.7 Map of solifluction lobes. Note relationship of lobe front to contours. (After Williams 1957) Profiles of lobe fronts indicated by symbols.

distribution of movement with depth varies. The advancing banks move over the ground surface downslope, and excavations often show buried plant remains that may extend many metres upslope beneath the advancing material (Fig. 11.7).

Although superficially similar features occur in other climates it is useful to reserve the name solifluction for features of this kind where freezing and thawing is an essential element in the mechanical processes involved. The movements occur on slopes which conventional methods of analysis (not taking into account the effects of freeze and thaw), show to be extremely stable. The relatively slow soil movement differs of course from the more rapid type of landslide movement that the engineer has long studied.

There are several effects involved. If the soil on the slope is substantially heaved, this can be

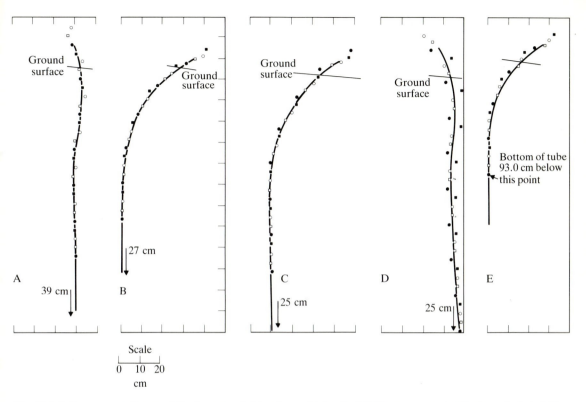

Fig. 11.8 Soil movement due to solifluction, revealed by shape of tubes buried three years previously, in vertical straight position. (After Williams 1966)

assumed to have occurred in a direction normal to the slope. The consolidation occurring on thaw takes place under gravitational forces acting vertically. The net result of heave followed by thaw may include a component of downslope displacement in the manner indicated by Fig. 11.9. However, the strength of the soil will be sufficient to at least partially resist settlement in an absolutely vertical fashion (studied comprehensively by Washburn, 1967). In fact, this particular mechanism of displacement (the earliest-suggested for downslope movement due to cyclic freezing and thawing) is probably of most importance in the immediate surface layer, and with reference to 'detached' particles or pebbles. The latter are often lifted on needle ice crystals – frequently seen on the soil surface in regions experiencing sharp frost.

Probably more important is the loss of strength associated with the loose state of frost-

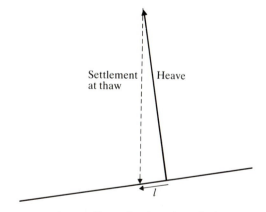

Fig. 11.9 Diagram illustrating downslope displacement following heave and vertical consolidation

heaved soil at thaw, in which the particles have been separated by the expensive process of ice formation. The cohesive strength of newly-thawed soil is as a result effectively zero. A commonly-invoked process is that of the 'supersaturation' of the soil at thaw but this explanation may be superficial. As the excess ice thaws the associated volume decrease on the transition to water tends to lower the high pore water pressures otherwise to be expected with the abundant water. High pore water pressures would of course promote strength loss and movement. But the expanded state of the soil mineral matrix, coupled with rapid drainage, apparently produce, according to some observed cases at least, pore water pressure less than atmospheric (although these may nevertheless be the highest pore pressures during the year). With such pore water pressures movement would not be expected on low-angle (e.g. less than 8°) slopes. Yet solifluction also occurs on such slopes. Artesian pressure would be necessary to cause movement (that is a pressure corresponding to a column of water greater than the depth of soil at the point in question). Occasionally, water trapped between a surface frozen layer and deeper permafrost, may develop such pressures (Chandler 1972).

Certain forms of solifluction occur on slopes of only a few degrees and there must be aspects of their mechanical behaviour as yet unrevealed. It has not even been established, for the case of the particularly low-angle slopes, at what time of year the movements occur. Possibly they are in some cases the product of creep in the frozen state, a topic considered further in section 11.10. Possible displacement of pipelines, railways, etc. make the matter one of practical importance. On occasion, conspicuous surface forms such as banks, or terraces, are absent, even though soil movements of a few cm yr^{-1} are occurring.

11.8 Thaw landslides and other rapid mass-movements

There is a common form of *rapid* mass movement, which involves the flowing or slumping of saturated soil masses, with a shear surface on the top of the permafrost or on top of a residual body of frozen ground in the spring. This type of movement which occurs in association with relatively steep slopes, and thus relatively high shear stresses, has to some extent its counterpart in other climatic regions. Abundant moisture and high pore water pressures because of drainage blocked by an impermeable soil or rock, are a frequent cause of slides in general. In blocking drainage the freezing of the ground has a more indirect and less unique role in causing soil movements than it has in the case of solifluction. Although mass movements of a nature intermediate between solifluction and 'pore pressure failures' also occur, there is a significant distinction between the fairly well-understood, rapid thaw landslides or flows, and the puzzling slow solifluction movements.

Thaw landslides are abundant for example, along many Arctic rivers, in deltaic or other low-lying areas, particularly where the base of the bank is being eroded. The bare frozen soil revealed by a slide is subject to accelerated thaw, and when ice-rich is prone to further sliding. The actual onset of sliding, and its frequency of occurrence is likely to depend on rate of thaw, and abundance of moisture. Thus it will be more frequent in warm, wet periods, when melt water tends to accumulate. Currently such slides are the subject of studies prompted by geotechnical considerations (McRoberts and Morgenstern 1974a). In some cases, part of the shear may simultaneously involve frozen soil (McRoberts and Morgenstern 1974b).

A variety of other forms of more or less rapid mass movement occur in association with snow and snow-melt water. Although avalanches are normally considered as involving snow alone, they often carry quantities of soil. When the snow is essentially in a melting state, slush avalanches occur. These consist often of mineral soil and snow carried in water originating in melt-water streams.

For the periglacial regions, attention often focuses on phenomena involving freezing and thawing, but Rapp (1960) compared the effects (quantities and distance of soil moved) of all mass movement processes in a particular periglacial area. He found that summer rains caused erosion, gullying etc. which were major elements. The most significant category of soil movement was 'earth slides'. He found that solifluction on the contrary, is a minor component in denudation, although the opposite view prevailed earlier.

11.9 Thaw settlement; thermokarst

A soil which has undergone frost heave looses volume during thaw by loss of that excess water, which was drawn into the soil during the freezing process. There is, in addition, a volume loss associated with thawing of any ice masses of different origin which may also be present.

The question of settlement due to thaw is complicated, and involves several stages. The transition of ice to water involves a volume decrease of 9 per cent, and if the soil is saturated, the soil should decrease its bulk volume accordingly. Such volume changes occur even in frozen soils (i.e. at less than 0 °C), when a temperature rise occurs causing the proportions of ice and water to change (Williams 1976).

A far larger volume loss is usually that caused by drainage, especially at completion of thawing (at 0 °C). Ice masses larger than pore size thaw only at about 0 °C, and as they do so the excess water will come to carry some part of the total stress or overburden pressure, σ. Drainage is accompanied by slow settlement. The ultimate amount of settlement depends on the static overburden pressure and the final pore water pressure, as these determine the effective stress. For near surface layers the component of effective stress σ' due to overburden pressure is necessarily quite small. Engineering activities (vehicular traffic, placing of foundations etc.) give large, sudden, short-term loadings, causing rapid build up of pore pressures, and an interim state of very low strength because of low effective stress. If the

water can escape easily as in very porous soils vehicular loads may also produce compaction (that is, consolidation). More commonly there is lateral shear and displacement, giving, for example, rutted, muddy roads, which may eventually dry out to produce an extremely rough road. The volume decrease directly associated with thawing of ice additional to the moisture content present before freezing is the greater part of thaw settlement. However, a recently-thawed compressible soil commonly shows a greater coefficient of consolidation (volume change per unit increment of effective stress) than the same material which has not been frozen. Paradoxically, this is related to the freezing process which causes consolidation of the soil matrix, but which also separates soil layers by ice segregations. Consequently the post-thaw consolidation process is the closing of cracks and discontinuities until the soil is a more or less uniform, close-packed assemblage of the heavily preconsolidated matrix layers. A preconsolidation pressure is revealed by any compressible soil that has experienced freezing and thawing, and the magnitude will depend on the lowest temperature reached (cf. p. 53).

One exception to the processes described, is the case of soil which prior to freezing carried a very large effective stress (e.g. soil which froze under a glacier). On thawing, the glacier now being absent, there may be sufficient swelling of the matrix layers by absorption of water, to cause water to move into the thawed material. Under such conditions the soil may in fact show a small volume increase following thawing particularly if the ice content was low in the frozen state. Some expansion of the preconsolidated matrix layers will occur in all compressible soils on thawing, but the effect is commonly totally overshadowed by drainage of excess water.

Situations involved in thaw-settlement are considered in detail, from an engineering point of view, by Tsytovich (1975). Two further general points are that the *rate* of settlement depends substantially on the rate of thaw. Thus if the thaw is occurring as a result of a uniformly maintained elevation of ground surface temperature, then settlement will be more or less proportional, to $\sqrt{\text{time}}$, just as will the depth of thaw (p. 166). If the surface temperature increases, the settlement increases as depth of thaw. Secondly, newly thawed soils have a high permeability which may be anisotropic (greater in one direction than others). Quantitative predictions of rates of settlement must therefore consider in detail the soil composition, hydraulic properties, in situ stress conditions, and thermal conditions. While the precise analysis of soil behaviour on thawing is difficult, surface forms arising from that settlement are very easily recognised.

Apart from geotechnical problems such as the settlement of new roads in permafrost areas, where new surface conditions lead to a net annual ground heat intake (a positive heat balance) there are widespread and varied natural effects. Some years after new surface conditions are established a quasi-equilibrium mean ground temperature is again established. When this is above 0 °C the permafrost ultimately disappears. More often the active layer reaches a new and greater quasi-equilibrium thickness. A sufficiently drastic disturbance, such as total removal of vegetation can increase the thawed layer initially by several metres yr^{-1}, the rate depending on the ice content of the soil.

Settlement due to thawing of ice wedges gives characteristic troughs; pits and depressions are sites of other ice bodies, and thaw ponds or lakes may ultimately develop. Equally important is the train of events that often follows such direct effects. Troughs and depressions may become water courses. The flowing water, because of its contained heat and good thermal contact with the ground, deepens the thawed layer and mechanically erodes sediments. Direct removal of ice by the melting action of flowing water is called thermal erosion (Mackay 1970) and is important where ice is exposed in river or stream banks.

Topography so formed is called thermokarst. French (1974) describes a 'badland' topography developed by thermokarst processes on Banks Island (72 °N). Slope retreat at rates up to 8 m yr^{-1} occurs by slumping associated with melting ground ice, and water courses have become wide gully-like features separated by residual mounds and cliffs. This remarkable example occurs in an area where apparently, the last period of climatic amelioration occurred thousands of years ago. Deepening of the active layer by climatic change seems unlikely to be solely responsible. Although much of the ground surface is in an almost continual state of movement it is not entirely clear what surface microclimatic characteristics changed to initiate this state of affairs. French (1976) considers it a result of a prolonged enlargement of ice wedges with consequent surface disturbance. The extraordinarily rapid denudation rates and associated land forms are, however, hardly typical of present-day permafrost regions. Their occurrence may also be related to the absence of climatic conditions for larger plants capable of stabilizing the slopes.

The Banks Island forms are an example of relief in which there are dominantly thermokarst effects on slopes and cliffs, and which consequently involves 'backwearing' and the lateral retreat of slopes. A distinction may be made between effects on slopes, and thermokarst relief mainly involving subsidence in the vertical direction, rather than downslope movement and removal. Such relief is widespread in ice-rich lowlands, and is characterized by formation of *alases*. In terrain with abundant ice wedges, deepening of the active layer (following from climatic change, or surface modifications) results in more pronounced troughs above the wedges (Fig. 11.10). Gradual slumping into the troughs occurs and standing water is abundant. As the sheets of standing water coalesce, there is additional thawing of permafrost, and settlement, because of the warmer conditions below the water. There is then further slumping of material into the emergent lakes – the alases. The alas lakes may migrate however, through the erosive action of waves against the shore, in the dominant wind direction. Such migration can lead to redevelopment of permafrost and even pingo formation in the formerly lake-covered ground. Thus there is a cyclic pattern of development of alases. Alas topography several thousands of years old as well as that currently forming can be recognized.

Apart from the 'ideal' sequence of events described, large areas of ground in Siberia, Alaska and Northern Canada, have an uneven relief, and much standing water, ascribable to

similar processes. Land clearance for agriculture is one of the many 'technical' procedures constituting ground disturbance that man initiates, and which can lead to thermokarst development.

11.10 Deformation in the frozen state

Until recently the possibility of downslope movements of frozen soil was almost totally overlooked. Even now, with the exception of those relating to rock glaciers, there are almost no field observations of downslope movement under natural conditions unequivocally due to deformation in the frozen state. In the meanwhile laboratory investigations, prompted mainly by geotechnical considerations, have revealed the special nature of deformation in frozen ground.

The effect of applying different static loads ($\sigma_1 < \sigma_2 < \sigma_3$ etc.) to frozen ground is shown in Fig. 11.11a, where strain is shown as a function of time. The condition of accelerating strain constitutes failure, as in the case of loads, σ_3 to σ_6. Loads σ_1 and σ_2 are carried indefinitely, at least if the temperature does not

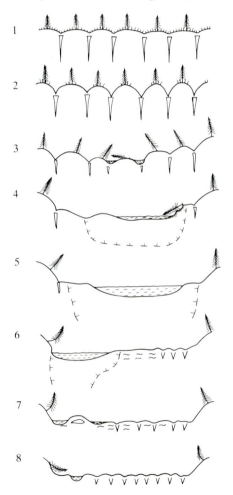

Fig. 11.10 Thermokarst: sequence of development and disappearance of alases. 1. Original lowland surface with ice wedges; 2. Effect of increasing depth of active layer; 3. Initial thermokarst stage; 4. A young alas; 5. Mature alas; 6. Old alas; 7. Stage of possible pingo formation; 8. Thawed pingo. (From Williams 1979)

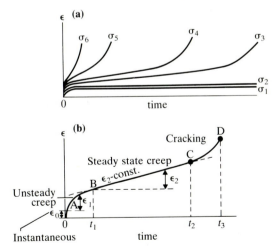

Fig. 11.11 (a) Graphs showing strain of frozen soil with time, for various applied loads σ, σ_2, *etc.*
(b) Curve showing the different stages of deformation. ϵ = strain; portion ϵ_2 shows constant *rate* of strain.

rise. It is most important that there is a range of loads producing failure, depending on the duration of application. In other words, while rather high stress is required to produce failure in a short time, a much smaller load perhaps only one-twelfth of that, is required to produce failure after a long period of time (Fig. 11.12). This leads to the concept of time-dependent strength, of 'long term' strength and of 'immediate' or instantaneous strength (Vialov 1965). Similar time-dependent behaviour is exhibited by some clays when unfrozen although the range of failure loads is much less. The behaviour relates to creep and stress relaxation within the material. 'Creep' is used here in a rheological sense, of continuing flow under loads lower than the strength with respect to immediate 'failure'. Stress relaxation refers to the declining resistance exerted by the soil in reaction to applied stresses. In frozen soils the phenomena are ascribable to the properties of ice and granular materials, and to the changing proportions of ice and water under applied stress or temperature change.

When a load is applied to frozen ground, there is a small, immediate, reversible elastic strain. There can be little freezing or thawing (necessitating heat flows) in such a short space of time, and to that extent the proportions of

ice and water and their disposition in the frozen ground do not change greatly. Such changes do occur subsequently over a period of time and are partly responsible for the subsequent deformation behaviour (Fig. 11.11a).

There is a tendency to regard frozen soil as merely combining the properties of ice and soil. It is valuable to compare and contrast the materials. Experiments show (Tsytovich 1975) that, provided the temperature remains constant, after the initial elastic strain frozen soils exhibit a viscous type of behaviour, with a strain rate dependent on and increasing exponentially with the stress. Unlike ice however, which continues to deform slowly under very low stress, generally a certain threshold stress value must be exceeded for the 'viscous' flow stage to occur in frozen soil. In this respect, frozen ground is a plastic material. The 'viscosity' of frozen ground is, however, commonly less, implying that flow is faster, than in pure ice. Whether or not the third stage with accelerating strain (σ_3 to σ_6, Fig. 11.11a and b) culminating in rapid plastic or brittle failure is reached, depends on whether or not the stress exceeds the 'long-term' strength. The strength decrease associated with duration of stress application is generally more or less complete after a few days or weeks. It is greatest during an initial period of hours or

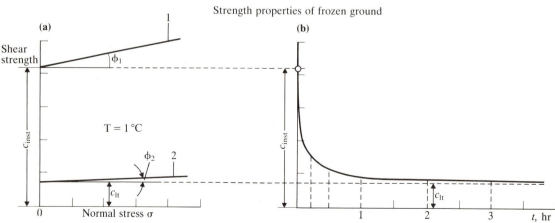

Fig. 11.12 Shear strength of frozen soil as a function of duration of load application. (a) Shear diagram showing values of cohesion, c, and friction angle ϕ, and shear strength as a function of a normal stress p. Values for 'instantaneous' (inst.) loading to failure, and 'continuous' or 'long term' (lt) strength are shown. In (b), the value of cohesion is shown as a function of time to failure. This represents the relaxation of cohesive forces with time. (Modified after Tsytovich 1973)

days. Some typical values of strength at –10 °C are shown in Table 11.1. The strength is highly dependent on temperature and on soil mineral composition. These strengths (Table 11.1) are very high in relation to stresses occurring in natural slopes, so that deformation in nature must, in general, involve different conditions. Notably, loading is normally 'long-term'.

The rheological behaviour of a particular soil, is dependent on the amount of ice (especially ice segregations) present. At one end of the scale frozen ground is an ice-rich material, which is essentially some soil material suspended in a 'glacier', at the other a relatively ice-free material in which soil particle contacts assume greater importance.

The shear strength, whether long-term, or instantaneous, is generally substantially greater than for the same material in the unfrozen state. This increase of strength on freezing is mainly due to increased cohesion provided by the ice. However, the friction angle in some soils may double between –1 ° and –2 °C, while the total strength (including cohesion) may double between –5 ° and –20 °C. Change of strength with temperature is greatest in the first few degrees below 0 °C. Figures from Vialov (1965), show the cohesion component of long-term strength (24 hr application of load) of 'sandy-loam' at −5 °C to be 44 times that in the unfrozen

state, while at −20 °C it is 97 times. A clay however showed cohesion only 1 to 2 times greater than in the unfrozen state. The picture is further complicated by the fact that the angle of friction of frozen soil varies with temperature, duration of loading (Fig. 11.12), and with normal stress. Observations of Vialov (1965) show, for a sandy loam at –10 °C, a friction angle of 29 ° (failure occurring after 1 hr. of load application) with normal stress between 0 and 20 10^5 N m^{-2}. With the normal stress between 20 and 50 10^5 N m^{-2} (again, for failure induced by 1 hr. load application), the friction angle was 19.5 °. The frictional strength actually developed, is of course, the product of normal stress and the friction angle, and must involve particle contact as in unfrozen soil. Presumably the high suctions developed in the unfrozen water of frozen soils produce a significant component of effective normal stress, and are thus significant in the frictional strength. The uncertainty surrounding the magnitude of the frictional component of strength is to some extent overshadowed by the much greater importance of cohesion.

The behaviour of ice under stress (Paterson 1969) is related to its crystal size and orientation. Although deformation of frozen ground commonly involves deformation of the ice phase *per se*, thawing of ice followed by

Table 11.1 Strength of frozen soils at −10 °C

Soil	Shear strength, 10^5 N m^{-2}	Time to failure	Normal stress σ
Callovian sandy loam	31.2	2 min	10
Callovian sandy loam	36.2	2 min	20
Callovian sandy loam	23.1	30 min	10
Callovian sandy loam	27.8	30 min	20
Callovian sandy loam	17.0	24 hr	10
Callovian sandy loam	21.0	24 hr	20
Bat-baioss clay, shearing surface			
perpendicular to laminations	23.0	2 min	10
perpendicular to laminations	25.2	2 min	20
perpendicular to laminations	12.6	30 min	10
perpendicular to laminations	14.8	30 min	20
perpendicular to laminations	12.0	24 hr	10
perpendicular to laminations	13.8	24 hr	20
parallel to laminations	17.0	2 min	10

From Vialov (1965)

migration of water and refreezing in a region of lower stress also constitutes an important mechanism for displacement of ice masses. While the strength of soils when frozen, measured over a short time, is far greater, often by orders of magnitude, than in the unfrozen state, the long-term strength at temperatures within a degree or so of 0 °C may sometimes be similar to that in the unfrozen state.

Under conditions of *changing* temperature, near 0 °C, additional considerations relating to creep are involved. A temperature change from –2 °C to –1 °C, for example produces a highly significant increase in unfrozen water content in many soils, the effect being even much more marked at temperatures above –1 °C. Associated with such changes are volume changes of several per cent of the bulk soil (Williams 1976). Cyclic temperature changes produce effects substantially in excess of those due to thermal expansion (without phase change) and which are invoked as a creep mechanism (Scheidegger 1970) in unfrozen materials.

The migrations of water and redistribution of ice considered in section 11.4 will occur primarily in the direction of temperature gradients with only a minor downslope component. Nevertheless, continual volume changes associated with temperature change, and the concomitant moisture migration and relocation of ice, seem likely to result in a certain net downslope movement of soil material. It seems probably that such movement would require only quite small shear stresses and thus low slope angles. Deformation in the frozen state in this manner may possibly explain certain forms of solifluction, which occur on slopes of low angle.

The picture gained from laboratory studies is one of complexity leading to an evident need for more information, but field evidence of deformations in frozen ground under natural conditions is also sparse. Movements of 0.2 to 1.6 m yr^{-1} have been recorded with surface markers on rock glaciers. From the surface and in aerial photographs rock glaciers (Pl. 11.1a

and b) appear as accumulations of boulders with a topographic form suggesting flow and often resembling a valley glacier (Wahrhaftig and Cox 1959). The presence of ice at depth has been established in a number of cases, although its origin, whether the ice is from ground water, buried glacial ice, or buried snow, is often uncertain. It can be concluded that the frozen state is essential for movement to occur in that the boulder-rich material would otherwise be quite stable at the relative low slope angles often found. Rock glaciers are not easily defined precisely (Johnson 1973) in so far as there are other boulder-rich slopes, sometimes parts of moraine ridges, which may be subject to similar movements. Because of the thickness and slope of the rock glaciers, the maximum shear stresses (occurring at depth, cf [2.1]) are 1 to 2 10^{-5} N m^{-2}. It is not clear however, what minimum shear stress is necessary to maintain the movement. If the strength of the frozen material is mainly cohesional then, just as in slopes of unfrozen material (p. 131), the maximum stable angle is dependent on the depth of the material and the height of the slope. Unlike frictional strength, cohesional strength does not increase with depth. The stability of frozen slopes ('stability' being regarded as the absence of all kinds of deformation including slow creep) is as noted further complicated in this respect by the effects of temperature change, occurring annually near the surface, and at greater depths over longer times. It appears that relatively warm, ice-rich fine-grained material at depth (and with mainly cohesional strength) may be liable to slow deformation even when the slope angle is quite small.

Since the role of deformation in the frozen state is largely speculation so far as field evidence is concerned, we shall also mention here 'cryoturbation' phenomena. Excavation in permafrost frequently shows a flow-like patterning in the soil profile. There may be orientation of stones and boulders with their long axes more or less parallel, and soil colouration as well as grain size differences giving an appearance of slow 'turbulent' flow.

(a)

(b)

Plate 11.1 (a,b) Rock glacier, Yukon Territory, Canada. These striking features are formed by slow movement, following from the presence of a core of frozen, probably ice-rich ground. The deformation properties of frozen ground seem the only explanation for the movements of a metre or so per year, which are in sharp distinction to the rapid movements associated with landslides. (Photographs courtesy of W. G. Nickling.)

Cryoturbations and 'involutions' are also very common in mid-European sediments and tills – in regions which had permafrost during the Pleistocene glaciation. Undoubtedly this patterning is partly due to soil movements, solifluction, slumping etc. in the thawed state; but it may also be that they are in part the results of prolonged deformation of permafrost materials. Clearly also an element in their origin is the differential expansions due to frost heave, that occur locally in association with surface and soil variations as discussed earlier.

12

Dryness and high temperatures

12.1 Extent of the effects: dryness

The previous chapter dealt with the effects of low temperature as characterized by periglacial terrain, which although predominantly polar, or near-polar, also occurs in special situations, such as high altitudes in the low latitudes. The topics of the present chapter, similarly, are not exclusively relevant to the tropical low-latitude or desert areas, with which the climatic parameters of warmth, or lack of precipitation, are immediately associated.

The definition of arid or semi-arid have been the subject of much academic discussion. The most obvious approach would be through the small amount of precipitation or its irregularity. A modern view, however, associates aridity with excess of evaporation over precipitation. Large areas of the world have low precipitation, yet have indisputably wet ground conditions. This is particularly so in the circumpolar region. Northern Alaska has some 20 cm mean annual precipitation, an amount similar to that in the Sahara Desert, and less than half that of Tunis in so-called 'semi-arid' North Africa. The arid zone of Australia has 75 cm annual precipitation. The Alaskan north slope (the north coastal region adjacent to the Brooks Range) is an area of innumerable bogs and lakes, and a tendency to thermokarst development. Evaporation, however, is low because of the generally low atmospheric temperatures. An additional factor is the restriction of infiltration by underlying permafrost. In the Antarctic on the other hand there are some areas as noted previously (p. 171), where truly dry conditions do prevail in the ground. Similarly truly dry conditions occur in large areas in the middle latitude, especially in Asia (the Caspian Sea region, Mongolia and North West China). Monod (1973) presents a well illustrated review of the deserts of the world.

Comparison of precipitation with evaporation raises other questions. Over large areas of the world's surface, there is a rainy season – during which precipitation exceeds evaporation – and a dry season when the reverse is true. The components of the energy exchange change in magnitude (as discussed in Chapter 3), but if over a period of weeks or months there is a net excess of evaporation over precipitation, the dryness is usually obvious in the soil and vegetation. There are conspicuous and characteristic features in landscapes associated with such 'semi'-aridity, compared with more fully arid situations (see e.g. Petrov 1976). Differences in the distribution of rainfall within the year, or irregularity over a period of years also produce distinctive natural terrains. Savannah grasslands occur where the warm season is wet, and Mediterranean conditions when the winter is wet. Steppe grasslands are another example of distinct soils and vegetation associated with semi-aridity. But there are also certain properties and behaviour of soils considered on a much more local and specific basis, which are associated with dryness, and common to all the regions described. Even during short periods of dryness these become evident in all the environments considered, and indeed on occasion in the more temperature regions. Wind erosion, for example, can become a dominant process of mechanical erosion of the soil surface, at least occasionally, over much of the earth's surface. The thermal properties of soils, insofar as they depend on moisture content, are quite different in the dry state than when damp or wet; and, even in humid climates, soils are on occasion 'dry' in this respect. The effects of weathering, on the other hand, are more strongly influenced by a long-term pattern of dryness and the extent to which it is interspersed with wet periods. This is true too of denudation processes, where long-term dominance of particular mechanisms of erosion is apparent in the nature of the landforms. Thus we can distinguish between the immediate direct effects of dryness, and those which are more direct and cumulative.

The irregular nature of rainfall in the truly arid regions often appears to be as important as the overall small amounts of precipitation. Though there are few places totally lacking precipitation for many years, the absence of

regular annual precipitation means that vegetation barely exists, and those weathering processes which are associated with organic residues are essentially absent. The importance of water in the chemical reactions of weathering has been stressed, and the slowness of chemical weathering with the absence of many weathering products commonly found elsewhere, results from the lack of water.

As in detailed considerations of behaviour of the ground in other climatic regions, so also a fruitful approach to the study of aridity lies in detailed study and measurement of ground conditions and energy exchanges. The U.S. soil taxonomy system (see Sanchez 1976) defines soil aridity through such an approach, as existing when the suction of the soil moisture for a near-surface layer is greater than $15 \ 10^5$ N m^{-2} for more than 180 days in the year. This potential (relative to atmospheric pressure) is the long-accepted value at which wilting of plants is regarded as occurring. While this is a rational definition tied specifically to soil water behaviour, it may conflict with aridity defined in terms of atmospheric climatic parameters. However, the nature of the soil and its water retention capacity is of fundamental importance in the plant environment, and the soil water suction is the most direct indicator of the need for irrigation.

The decisive factor, in the energy exchange, with respect to evaporation rates in low-latitude deserts is normally the net solar radiation. It is this which produces appropriately high temperatures favouring evaporation. The dryness (low relative humidity) of the air in such deserts is also important. This low humidity together with the absence of cloud cover (both are at once a result and a cause of the arid conditions) enhances evaporation by allowing a high percentage of solar radiation to reach the ground surface. Nevertheless, desert regions are not always regions of low or sparse cloud cover, and low relative humidity. The coastal deserts on the west side of the continents of Africa, America and Australia are characterized by cloudy conditions and relatively high humidity of the air. The Equatorial coastal desert is marked by constant cloud cover and a high humidity going over into drizzle. These deserts however have very low precipitation, which must be regarded as the main climatic factor in their formation.

In summary, the concept of desert is complex, and involves an extreme or dominating role of effects associated with dryness. Such effects, often precisely because periods of wetness intervene, take many forms, and are represented in some degree in most parts of the world. The more extreme consequences of dryness, and the spread of these – 'desertification' – have great importance for human well-being in many of the poorest parts of the world. Largely prompted by famines, or by the technological needs of developing countries, the scientific study of the effects of dryness on the earth's surface is increasing, although it is still far behind the study of earth materials in more temperate conditions.

12.2 Extent of the effects: high temperatures

Although high temperatures promote dryness, they may also be associated with particularly wet conditions, as in the humid tropics. The rainfall is, of course, related to patterns of atmospheric circulation, just as the absence of rainfall in the desert regions adjacent to the tropical humid belts is also ultimately an effect of global atmospheric circulation patterns. The effects of wetness in association with high temperatures, are to produce an abundant and characteristic vegetation, and most notably an abundance of products of chemical weathering. So significant is the presence of water both for biological and chemical activity, that it is hard to find surface relief, soil profiles, and soil materials whose nature is ascribable to high temperatures, and which are common to the extreme arid and the humid conditions. Thus while dryness gives rise to characteristic

properties and behaviour of earth materials, the effects of high temperatures will always be greatly modified depending on the absence, or presence (or the relative amount) of water.

Thus, this chapter will be broadly divided into the topics of dryness (which may involve high temperature), and of high temperature associated with wetness. It will also be noted that many of the effects of dryness are associated with alternation of dryness with wetness.

12.3 Weathering in hot, dry climates

Arid regions are characterized by very slow rates of weathering arising from the scarcity of water. Such rains as do occur usually penetrate only a distance of at most a few metres into the ground before moving upwards again to evaporate in the ensuing dry period. Characteristically, chemical reactions are slowed by the accumulation of weathering products (see p. 28) which are not removed in solution. This does not usually result in thick, homogeneous layers. Rather, unaltered parent material is abundant and interspersed with the primary weathering products. Soluble materials may, however, move upwards in ground water to be deposited when evaporation occurs. Particularly in low-lying regions serving as drainage basins, movements of water upwards from the ground water may lead to the deposition of salts on or just below the ground surface.

The small amounts of clay minerals which are formed tend to have a high concentration of adsorbed cations. An absence, or near absence of vegetation, and the rapid oxidation of organic remains at the relatively high temperatures, results in minimal organic residual compounds. The absence of such compounds further reduces the possibilities of chemical change. Conditions are highly oxidizing when temperatures are high and these are responsible for the reddish or yellowish colours of soils, which are usually due to ferric iron compounds. In colder deserts, oxidation is less pronounced, and the accumulation of some organic matter in the partly decomposed soil parent materials, may lead to products of a grey rather than reddish colour (Loughnan 1969). Notwithstanding the importance of all these general characteristics, there is much diversity in the weathering processes of arid and semi-arid regions (well-described in Petrov, 1976).

The effect of high temperature, or to be precise, of high temperatures alternating with much lower temperatures, is to promote physical weathering. Diurnal *air* temperature variations are not necessarily particularly large in deserts (contrary to widespread belief), but unimpeded solar radiation during the day brings the surface of the bare ground to a high temperature. The somewhat higher albedo of the soils compared to that of vegetation, is not sufficient to reduce the net solar radiation greatly, and the absence of heat consumption in evaporation also promotes a high surface temperature. It is particularly evident in sandy materials which, when dry, always have low heat capacity (and therefore warm rapidly) and very low thermal conductivity (see p. 63). Thus high temperatures occur in a shallow surface layer. During night-time outward long wave radiation to the clear sky of many desert regions is also high. Thus the diurnal temperature range of the ground surface is large, for example as much as 70 °C, but the reduction of the diurnal temperature wave with depth in the soil is often very abrupt. Dry soils have a low diffusivity (see Fig. 5.5) as a result of their very low conductivity, and the low diffusivity is responsible for the damping of the temperature wave with depth according to equation [10.1]. The same applies to the annual wave which Monod (1973) reports can be more or less absent at 1 m depth. This however, is also due to the small annual temperature amplitudes characteristic of the tropics (associated with the small annual variation of the sun's angle summer and winter) which are also to be expected for temperatures in the ground. At 50 cm depth in the tropics there is (Sanchez 1976) generally less than 5 °C variation, summer to winter. In continental deserts in *middle* latitudes the annual amplitude at the surface may be more than 120 °C (Petrov 1976), with a maximum temperature of 80 °C. The extent of temperature change diurnally, and the high temperature gradients in the near surface layers, must be important in creating stresses within rocks. At the same time the much higher diffusivity of continuous rock results in surface temperature variations in solid rock being transmitted to much greater depths than in soils.

Contraction of a rock on cooling can be compared to an equivalent compression, produced by an externally applied stress (see Lerman 1979). If the corresponding stress (to produce the same volume decrease as produced by the cooling) exceeds the tensile strength of the rock, rupture is likely as a result of the temperature effect. The extent of physical weathering in desert conditions and the effects of thermal contraction also often depend on the difference of coefficients of thermal expansion of the different mineral components of a rock or pebble. Over all, however, the disintegration of rocks and pebbles by thermal stresses is slow: experiments have shown many thousands of temperature cycles to be necessary to cause fractures (Journaux and Coutard 1974). Fatigue failures arising only after repeated stressing are probably important.

A special and a relatively important category of stresses associated with fracture and weathering, are those arising from various salts. Concentrations of salts, sometimes in commercially valuable quantities, are common in arid and semi-arid terrain. Some are weathering products *in situ*; but it is now recognized that the absence of sufficient rain to produce an equivalent amount of weathering, often means that the salts must have a distant origin. Yaron *et al.* (1963) believes that the abundant salts in soils in parts of Israel, are of marine origin and are borne by winds from the sea. Elsewhere, ground or surface water flowing substantial distances, may terminate in low-lying *playas* where there is steady evaporation and a consequent precipitation of salts. Such *evaporite* deposits are frequent in playas, which are defined as desert drainage basins (Cooke and Warren 1973), and are often hundreds of square kilometres in extent.

'Salt weathering' describes the weathering involving a progressive forcing open of small cracks and openings so that small or sometimes large rocks are ultimately shattered. The expansion of the salts may occur because of hydration, and this becomes a repetitive process where intermittent high temperatures and low humidity occur and produce the anhydrous state. Hydration can produce pressures sufficient to enlarge pre-existing cracks or openings, leading to rupture, particularly when the hydrating salt is confined in the furthest point of a crack. Marshall (1977, p. 64) cites a particularly interesting example which illustrates the importance of the Gibbs free energy in analyzing the process. In Chapter 2 it was pointed out that for a chemical change to occur, spontaneously, the free energy of the products must be lower than that of the reactants, so that there is a decrease of the free energy of the system. Goethite is the hydrated form of the ferric oxide hematite as illustrated in the equation:

$$2\ Fe\ O(OH) \rightleftharpoons Fe_2O_3 + H_2O + 0.8\ 10^3\ J$$

Goethite Hematite   Heat of reaction

There is a particular free energy of water below which the reaction occurs in the direction to form the hematite. This free energy corresponds to a suction in the soil water of $5\ 10^7\ N\ m^{-1}$. Equally, it corresponds to a relative humidity of the air (see p. 94) of 68 per cent. At lower suctions, or higher humidities, the free energy of the water is such that the reaction occurs in the opposite direction to produce goethite. The critical suction is of course associated with a rather dry state of the soil and one through which there may be repeated transitions, as moist and dry periods alternate. Consequently there will be repeated transitions between goethite and hematite, with the associated volume changes of the compounds. The simplicity of this example, linking the state of the soil water, or the relative humidity of the air, with the direction of chemical change, and the consequent weathering, is somewhat qualified. The size of the soil particles, the pores and the salt crystals are all small, which tends to modify the free energy of the other components of the reaction as well as that of the water. Nevertheless, even if the critical suction or relative humidity may accordingly differ from those given, the principle is well-illustrated.

The growth of crystals in confined spaces by precipitation from solution has similarities with ice lens formation in frost heave (p. 98), and is quite distinct from hydration. The confined solid-liquid interfaces are the cause of the elevated pressure generated within the growing crystals and exerted by them on their surroundings.

A third mechanism of stress generation involving salts is simply that of thermal expansion. According to Cooke and Smalley (1968) the volumetric thermal expansion of various, quite common salts increases unusually rapidly with temperature; while rocks, for example granite, do not show a similar increase (Fig. 12.1). The implication is that the large diurnal temperature changes in near surface materials in warm regions will be associated with the generation of substantial internal stresses when salts are confined within rocks.

A further form of mechanical weathering associated with dryness is that due to simple alternation of wetting and drying. It is well known that many cohesive soil materials and weaker sedimentary rocks are caused to flake by re-wetting from a dry state. The strong attraction of many porous dry materials for water implies that even re-wetting (considered on the scale of pore size) will result in considerable expansive stresses. Particularly where rewetting is not uniform there will be concentration of stresses leading to cracking

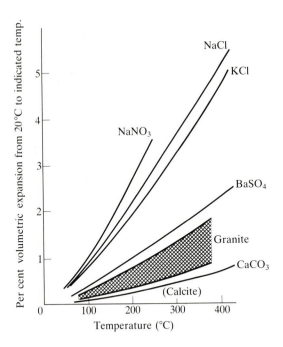

Fig. 12.1 Relation between temperature and the volumetric expansion of various salts and granite (it is assumed similar relationships apply at common atmospheric temperatures) (After Cooke and Warren 1973)

both on the micro- and macroscales. The concept of residual strain energy has been advanced (Chapter 9) to explain the fissuring and disruption of heavily over-consolidated clays on rewetting, and it is reasonable to assume that many sediments when 'dry' to the extent occurring in arid regions, will also have a similar residual strain energy. The process of drying with the development of suction causes a considerable consolidation which may involve an elastic deformation of particles and particle structures (Chapter 4).

It should be noted that although precipitation may be very infrequent, it is still commonly responsible for some development of soil horizons even in quite arid situations, with the water moving primary weathering products. Even sparse vegetation will provide some organic residues which will promote chemical change. Erosion and sedimentation by water often appear to be important as geomorphological agents in desert regions, and this is partly because rainfall, when it occurs, is concentrated, and also because of the absence of protective vegetation.

12.4 Weathering and soils in hot, humid climates

The ambient temperature may determine the *direction* (that is the eventual products) of chemical change in weathering and soil formation. The goethite-hematite alteration described in the previous section, is an example, because the free energy (and vapour pressure) of the soil water are temperature dependent. But it seems probable that, in general, the characteristics of weathering in the tropics, and the tropical soils, are more related to the *rate* and *extent* of the reactions. The effect of temperature on rate of reaction has been described in Chapter 2. The importance of water in removing the products of reaction, so as to allow reactions to continue beyond the point at which accumulating products would otherwise establish equilibrium with the reactants, was also noted.

Leaching by infiltrating precipitation is the dominant process in the formation of the residual soils of humid tropics. Loughnan (1969) after Strakhov (1967) considers the rates of removal by solution of silicate materials: the rates depend on infiltration and thus precipitation, and temperature. He concludes that weathering takes place between 20 and 40 times faster in the humid tropics than in the temperate regions. Certainly the humid tropics have great depths of weathered material, often many tens of metres, which, dominantly, has lost much silica and often consists largely of alumina and ferric oxides or ferrous oxides. Rain water flows downwards through the profile to the water table (which is often close to the surface), and continuing ground water flows ensure the removal of soluble material. The hydrated alumina and ferric oxides which are particularly prominent towards the ground surface, constitute bauxite and laterite. The term laterite is often used in a general sense to describe the soils of moist, tropical regions. But, notwithstanding the dominance of the moisture and temperature conditions, soils and weathered materials of the warm regions show as many variations as those of the temperate regions. There is a spectrum of soils, Fig. 12.2 (see Sanchez 1976), from those associated with characteristically arid conditions (section 12.3) to the extremely weathered materials described above in relation to humid tropics. The climatically-induced characteristics are superimposed on the mineral and lithological characteristics of the parent material.

A general point of contrast with the temperature regions needs emphasis. In moist, humid, warm regions, the depths of chemical weathering are so great that geotechnical engineers will normally be using such weathered materials in foundations. In temperate regions the limited vertical extent of weathered soil horizons means that their chemical characteristics are quite subordinate to the physical properties of the parent sedimentary material – to which engineers extend the name soil. Gidigasu (1976) points out the inappropriateness of parameters

derived from temperate land practice, especially those of grain-size composition and plasticity, as a guide to the geotechnical and mechanical properties of tropical soils. The correct approach to soil classification and description for geotechnical purposes, he believes, is through consideration of the tropical soil-forming processes, that is, the study of *pedogenesis*. These processes are of course, primarily chemical change (chemical weathering). The subject is normally ignored in geotechnical education.

Humid, tropical environments are normally associated, in the mind's eye, with profuse vegetation. The nature and amount of vegetation may control weathering to a large extent (Tricart 1972). Modification of the micro-climate by shading and protection, maintaining uniform conditions of moisture and temperature, limits physical weathering. In Guinea a diurnal amplitude of only 1.2 °C was observed 20 cm below the surface in a forest, while under exposed ground with a hard crust the amplitude was 13.6 °C (Tricart 1972, after M. Aubert).

Vegetation and its remains produce intense

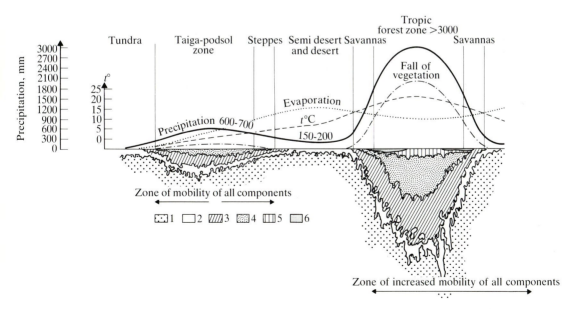

Fig. 12.2 Diagram illustrating weathering of surface mantle under various climatic and vegetation conditions. 1. Fresh rock;   2. Zone of gruss eluvium, little altered chemically;   3. Hydromica-montmorillonite-beidellite zone;   4. Kaolinite zone;   5. Al_2O_3;   6. Soil armour, $Fe_2O_3 + Al_2O_3$. (After Strakhov 1967)

biochemical weathering in warm regions: humus in the topsoil causes soil aggregation, and a permeability generally sufficient to ensure infiltration rather than run-off. Infiltration in turn ensures optimum conditions for leaching. Carbonation, the effect of carbon dioxide released from rotting vegetation and dissolved in water, may be important. Also, relatively more intense leaching occurs at 0.5 to 2 m depth, causing a concentration of plant roots near the surface where there is ample supply of nutrients from litter. The chemical weathering residues together with the nature of the parent materials, give rise to soil water characteristic curves (suction-moisture content relationships) which are often distinct from those associated with the (unweathered) soils of temperate regions (Fig. 12.3). These curves reveal the cause for the various reactions of soil to drying conditions; the reactions are important for agriculture. The sandy soil of Fig. 12.3, is characterized by the small amount of water held in the suction range above $0.1 \ 10^5$ N m^{-2} (= 10^2 cm H$_2$O). Similarly, the clayey oxisol has a relatively small amount of

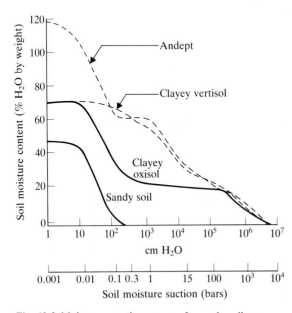

Fig. 12.3 Moisture retention curves of a sandy soil, a clayey Oxisol, a clayey Vertisol, and an Andept. Andepts and vertisols are soils of the humid tropics. (After Sanchez 1976)

water available in that suction range. Plants are likely to suffer distress sooner in such soils than in the case of 'andept' soil or 'clayey vertisol' where large amounts of moisture remain available to the plant, even when the environmental conditions have produced soil suctions between $0.1 \ 10^5$ N m^{-2} and the permanent-wilting point (see Baver, Gardner and Gardner 1972) of $15 \ 10^5$ N m^{-2}. The latter suction represents a dryness of the soil at which most plants can no longer extract water (their potential being too high) and consequently wilt.

Just as truly dry areas are of limited extent, so many humid areas are liable to be affected by local, temporary dryness. An example is the changes that often occur as an attempt is made to practice intensive agriculture on the site of cleared forest. The reduction in vegetable remains is aided by their rapid oxidation, promoted by the dry, hotter and more changing conditions of the exposed surface. Fertility decreases and the aggregate structure of the soil is lost, and with the decreased permeability there is less infiltration. Instead heavy storms will lead to run-off and water erosion of the surface may occur. A brick-like crust may develop. If the exposed soil is cracked, the cracks may channel water to become the site of gulley erosion.

In sum, the fully humid tropics are characterized by the rapid rate of weathering associated with high temperature and abundant moisture, which in turn explains the often great thickness of weathered material. Sharp transitions in thermal and moisture conditions with their effects, are also characteristic. Local or periodic transitions to much drier conditions are important. Such is the intensity of net solar radiation and the ensuing rate of evaporation in exposed surfaces, that even a week without rain, during the rainy season, may cause a wilting of the crop (Sanchez 1976). These briefly-outlined characteristics of the humid tropical areas, and the arid areas described previously, represent the challenge to overcoming the famines and low living standards generally of those regions. These challenges must be met by a scientifically-based approach to agricultural, hydrological and geotechnical problems.

12.5 Particle sorting and volume change effects; ground patterning

Soils in the tropics exhibit a variety of micro-relief and ground patterns, many of which bear some resemblance to the patterned ground and other features so characteristic of periglacial regions. Different types of these features occur in humid and arid situations although almost without exception water plays some role. The alternation of dryness and wetting of the soil is especially important, just as thawing interspersed with freezing is essential for the formation of many of the periglacial features.

The general term *gilgai* covers a range of patterns – cracks, mounds, depressions and step-like forms, in which the swelling and contraction of soils rich in the 'swelling' montmorillonite clay minerals is important. Gilgai features are widespread throughout warm regions of the world where there are alternating wet and dry seasons. Vertical cracks in clay-rich soils opened during the dry season, become partially infilled by near surface material falling in, the process being enhanced by rain. Moisture causes swelling and closing of the cracks and compressive stresses arise particularly because of the trapped material. These stresses may cause upheaval of the surface, between the cracks, and numerous shear planes at some metres depth. The surface may be further modified by wind erosion.

The gilgai phenomena are also modified by the nature of the soil. Somewhat coarse-grained material (which is not subject to marked volume change) may show only alluviation of clay particles which are carried down through the soil pores during the rainy period. MacFadyen (1950) describes 'vegetation arcs', in Somalia which are narrow belts of grass and scattered trees, essentially paralleling the contours of slight slopes at some 160 m

intervals. Each vegetation arc, he suggested, marked the sight of deposition of organic material by an ephemeral body of water. Hemming (1965) proposed instead, that the arcs are the result of grazing, cutting or burning. Such activity would produce bare patches with trampling leading to lower permeability. Run-off from these segments of the slope would nourish the vegetation remaining, which in turn would capture more water, soil and organic materials. Features in which vegetation belts run downslope (i.e. normal to the contours) are also found. Associated with the vegetation are modifications in the near surface soils, which are probably a result of wind and water erosion (Cooke and Warren 1973).

Sub-surface erosion (piping, p. 144) by running water gives sub-surface openings and leads to surface subsidence, with development of features quite characteristic of some desert areas. Piping is associated with permeable materials in which a small component of fine-grained material passes through larger continuous pore openings.

Other subsidence features of dry areas are apparently the result of wetting and the ensuing packing or consolidation, of loose, weathered material. This phenomenon often occurs in association with irrigation, when it is reasonable to assume it is a consequence of saturation of materials not substantially wetted previously. Phenomena associated with salts as described in section 12.3, are also involved in the formation of various ground patterns.

12.6 Water and wind erosion

It may seem anomolous, in the context of aridity, that the effects of running water are often conspicuous in dry lands. But it is such effects which frequently produce the variety in the surface materials and landforms of arid and semi-arid regions. Water as an agent of erosion distinguishes terrestrial arid environments from those, for example, of the moon. The forming of slopes, of valleys and river beds by the action of water is fundamental.

The rate of overall denudation of arid lands is nevertheless much less than in other regions. Although the figure varies substantially, a rate of $0.5-1$ m^3 km^{-2} yr^{-1} may be typical for deserts (Cooke and Warren 1973).

The quantity may be hundreds of times as great in the humid tropics, 7 tons per hectare per year – according to Sanchez (1976) – that is, about 500 m^3 km^{-2} yr^{-1} – and a thousand times as great in, for example, humid, glaciated polar regions. Obviously the relative scarcity of water for removing sediments is the main factor.

The landforms of arid areas are not infrequently relict, that is they were formed during a period of greater rainfall probably thousands of years ago. Carson and Kirkby (1972) describe evidence of alternating periods of erosion and deposition that may be associated with alternations of wet and dry periods. The inference is that the greater erosion is associated with the drier periods when vegetation is sparse. This is seemingly in conflict with the general trend for greater erosion as rainfall and humidity increase. Because mass movements in general are infrequent (there being few landslides or mudflows) in arid or semi-arid lands, overland flow and erosion becomes relatively important.

In fact the role of plants in reducing erosion, especially water erosion, is very significant. Although precipitation, and thus overall sediment yields, are small in arid areas, the exposed and loose nature of soils free of vegetation, and the concentration of rainfall into short periods, means that is relatively effective in eroding surface materials. The nature of many soils in semi-arid regions especially, is such that the effect of raindrops is to form a hard-packed, relatively impermeable surface layer. This is also true of soils in humid tropical regions, such that clearing of forest is associated with rapid soil erosion. The absence of vegetation means that the impact forces of raindrops are high. Rainsplash becomes more

effective in loosening particles, as well as in promoting a caking of the surface.

Once overland flow has been established the amount of erosion will depend on velocity of water flow, and on the size of particles according to the classic relations described by Hjulstrom (1935). (Fig. 12.4). Very small particles are only loosened by rather fast-flowing water, because of the cohesive forces holding them to the soil matrix. Such particles on the other hand remain in suspension and are thus transported even at low water velocities.

Wind erosion, so characteristic of the arid or semi-arid regions, is of greatest economic significance when it occurs in more temperate, and otherwise fertile, regions – the 'dust bowl' of the thirties in North America being the prime example. Such extensive erosion, even though initiated by unusual weather conditions, is not likely to occur again in developed countries, because of the increased understanding of the phenomenon and the control which man can exert upon it (reviewed in Chepil and Woodruff 1963, Beasley 1972).

Wind erosion normally only produces conspicuous effects when it occurs at a rate exceeding that of soil formation. Lesser amounts of wind erosion may nevertheless have importance because of the loss of thickness of fertile soil. Wind erosion may be confined to small areas, even to particular topographic features or parts of them, or it may occur

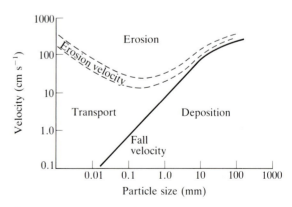

Fig. 12.4 Critical water velocities for erosion, transport and deposition as a function of particle size. (After Morgan 1976 from Hjulstrom 1935)

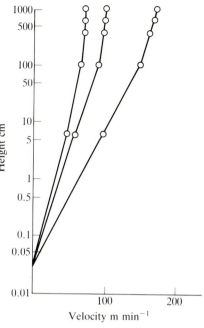
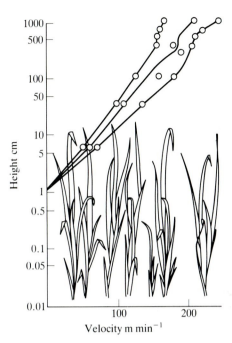

Fig. 12.5 Windspeed variation with height over a flat sandy surface, and a grassy field, showing the low velocities on vegetation covered surfaces. (After Chepil and Woodruff 1963, *in* Embleton and Thornes 1979)

essentially simultaneously over areas of tens or hundreds of square kilometres.

The basic causes of wind erosion are best understood by an initial consideration of the erosion processes. Wind exerts a force on particles by its impact. Wind also exerts a drag on particles of soil exposed to it because of the viscosity of the air passing around the mineral surfaces. A third effect follows from the wind velocity increasing with height above the ground surface. Where such a velocity gradient occurs, there will be a pressure gradient in which the pressure of air decreases as the wind velocity increases (this is often referred to as the Bernouilli effect). Consequently a particle will tend to rise, pushed from below into the lower pressure region above it, particularly if the particle is dislodged by the impact pressure of the wind, or indeed by the impact of another particle. Figure 12.5 illustrates the wind velocity gradients over a sand surface, and a vegetation-covered surface – where the susceptibility to erosion is much less.

The dislodging, lifting, and falling back, constitute the *saltation* of particles – which

move in the direction of the wind by a succession of such jumps (saltations). The saltating particles are some tenths of a mm in diameter. Particles also move rapidly in a near surface layer of air (a few mm thick), this constituting *surface creep* and involving particles 0.5 to 2.0 mm in diameter. While saltation lifts particles up to 30 cm above ground, small enough particles (less than 0.1 mm) may be carried in *suspension* to much greater heights. Suspended materials gives rise to dust storms.

Dislodgement of particles on the ground surface is, of course, resisted by the strength of the bond between the particle and the substrate. This is usually a cohesion bond, often associated with adsorbed water, and perhaps with cementing materials. Cohesion is associated with small particles (Chapters 2 and 9) and accordingly very small particles will be less easily dislodged than somewhat larger ones, as shown in Fig. 12.6. As the size of particles, and their density, increases, as would be expected, greater wind velocities are necessary to initiate erosion. In Fig. 12.6, the

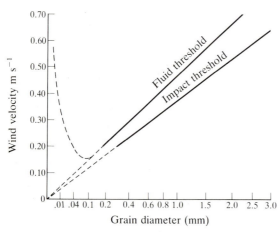

Fig. 12.6 Critical wind velocities for erosion as a function of particle size and density. (After Cooke and Warren 1973)

threshold wind velocities are shown as a function of the diameter of the particle. When particles are already in suspension, a lower velocity (impact velocity) maintains the erosion.

Although the drag effect of the air is greatest over a rough surface, erosion may be reduced if the projections (pebbles etc.) on the surface are relatively inerodible. Surface projections then served to protect low-lying parts of the surface which trap smaller particles. On a larger scale, ridges and knolls or other exposed topographic features are the first to be eroded.

The tendency for particle displacement and transport by wind is much influenced by the properties of the soil and the microclimatic or other characteristics of the ground surface region. Vegetation reduces windspeed. Roots strengthen soil. Thus erosion by wind often follows loss of vegetation. Vegetation may also serve as a windbreak reducing erosion down-wind. The longer the unimpeded flow of air, the greater is the amount of erosion.

Dry soils are more exposed to erosion, dampness giving either cohesive strength to the soil, or, if wet enough, a strength effect due to negative poor water pressure arising from capillarity. The nature of the soil, the grain size, density of component minerals, and soil structure are important. The soil structure, the aggregation of particles into granules or clods, depends on the organic content as well as that

of various cations and cementing agents. Aggregations are resistant to wind erosion because of their size and strength. Many of the factors responsible for the soil structure (water-soluble salts, soil colloids, organic products of decomposition) may directly reduce erodibility of a soil surface, and thus provide resistance against breakdown of the aggregate itself. They may lead to a hard, relatively impervious crust, significantly resistant to wind erosion. Fine sand particles, having no cohesive attachment, are easily dislodged and moved by wind, but clay and silt particles bind the soil.

Whether or not wind erosion occurs usually depends more on the nature of the ground surface than on the wind itself, although of course, the timing and extent of eventual erosion will be related to the wind occurrence.

The factors described are much influenced by human activity. Loss of vegetation follows overgrazing or burning, or may simply result from clearing for agriculture. Soil structure is liable to much change as a result of cropping practices, irrigation, fertilizing, and other associated effects. Proper agricultural practice also involves windbreaks, and ploughing or planting in directions likely to reduce the risk of erosion. The stability (or instability) of the ground surface, its resistance to erosion, is obviously, to some degree, a function of the climatic conditions. But it appears to be the activities of man which most commonly *initiate* large-scale soil erosion by wind.

In those parts of the deserts of the world where the surface materials are loose sandy or silty material, the constant shifting of these materials by the wind constitutes the 'normal' state of affairs. Where agriculture or other human activity is disturbing the natural vegetational and other environmental conditions the prediction and mitigation of wind erosion is a major concern. Prediction is based on assessment of a range of factors which are then combined into an equation for potential soil loss per year, in a procedure similar to that for water erosion, discussed in Chapter 10. One such equation is (Chepil and Woodruff 1963):

$$E = f(I, C, K, L, V)$$

where I = soil erodibility
C = local wind erosion climatic factor
K = soil surface roughness
L = equivalent width of field (relating to unsheltered distance in direction of wind)
V = equivalent quantity of vegetation

The mathematical relationships between the factors in the equations are complex but various charts and tables are used to arrive at E. The values of I, C, K, L, and V are assessed according to various rules, for example, I is based on grain size, and C on special climatic maps of quantities important in wind erosion. Several of the factors are composite, being based on assessment of several different characteristics within the category represented. For example, the vegetation quantity includes assessment of amount and form.

Control of wind erosion is based on conservation and cropping practices which lead to ground surface conditions least liable to erosion (Beasley, 1972).

12.7 Desertification

The spread of desert conditions on originally fertile soils, the lack of vegetation and the shortage of moisture for re-establishing plant growth, with the ensuing famine and hardships for the human population, have been particularly prominent in the last decade. The recent publicity surrounding famines in Africa, the drought in the Sahelian desert, and the extension of the boundaries of the world's deserts, is a consequence of humanitarian concern and mass communication. Such current newsworthiness can lead to a misinterpretation of the conditions causing desertification. These conditions, to the extent that they are largely natural or semi-natural, should probably not be considered as abnormal, but merely as recurring at regular quite short, intervals. Carson and Kirkby

(1972) report evidence of alternations of period of fluvial erosion and sedimentation in semi-arid regions during the past century and longer; this corresponds to the alternation of vegetated and vegetation-free or sparsely vegetated land surfaces. It is well-established that movement of sediment and establishment of new drainage channels are enhanced when vegetation is poorly developed.

The basic cause of low rainfall and thus the presence of the desert belts of the world, is the subsidence of air masses, that is, the movement of air downwards in the atmosphere. These movements are associated with global air currents. Air moving downwards becomes warmer, and thus can hold more moisture. This is the reverse of raining – which is associated with the rising of air and condensation of water vapour, that is, precipitation. The spread of deserts however is another matter, and some authors (see e.g. Monod 1973, Glantz, 1977) rule out gross climatic change as the cause of recent 'desertification'. Desertification follows from changes of the conditions at the earth's surface. It is clear that desertification can never be wholly ascribed to change of the global climate. Quoting from UNESCO (1977, p. 11): 'To see precisely what happens when desertification occurs, attention should be focused on that shallow meeting place between soil and atmosphere, where plants thrive and where a balance is maintained between incoming and outgoing energy and between water received and lost.'

In examining the effects of different surfaces and their effect on the microclimate, the atmospheric climate and its variations should not be ignored. But rather than considering these variations as the cause of surface change, they should be viewed as providing part of an unstable environment. Drought will inevitably occur at frequent intervals, in parts of the world. In such an environment, changes of microclimate, which are often man-induced, may have particular effect. As Carson and Kirkby (1972 p. 349) put it: 'Under either fully arid, or humid conditions, vegetation cover is unaffected by small changes in external conditions, but in semi-arid conditions even very small changes in rainfall, human water use or grazing patterns may produce considerable changes in vegetation cover to which the landscape responds'. Several authors (e.g. Rapp *et al.*, 1976) note that soil erosion is often accelerated *after* a period of intensification of land use in marginal areas during wet years. Grazing may be increased or natural vegetation areas cleared for cropping, exposing the surface to erosion during the subsequent, perhaps more 'normal' dry years.

Loss of vegetation cover can double the albedo or reflectivity coefficient (compare Table 8.2), decreasing the amount of absorbed solar radiation. The entire energy balance is changed drastically, in a complicated way. Not least the ground surface temperature is changed although often not in the way that might be expected, and in association with this there are changes in all the fluxes at the surface. The nature of these effects is still a matter of dispute (see Hare in UNESCO, 1977, p. 99). If the higher albedo led to lower surface temperatures it would promote subsidence or stable air conditions, that is, lack of rain. On the other hand it is common knowledge that a bared ground or rock surface is usually warmer to the touch than one under vegetation; this is largely due to the lack of moisture in the exposed surface, and thus absence of cooling by evaporation. Discerning which is cause, and effect, is made more difficult by the dependence of albedo on the soil moisture content (Idso and Jackson 1975). Dry soils indeed have higher albedo than wet ones.

When ground is bare and dry, and the desertification process is initiated, dust is abundant in the air. The dust adsorbs solar radiation and in turn imparts heat to the atmosphere. Warming of the atmosphere above the ground in this way reduces the chances of warm air rising, and thus indeed decreases rainfall. This mechanism producing arid conditions, is an example of a microclimatic change causing at least local climatic change. At least this is so, if the drying of the surface

(leading to wind erosion, and following from loss of vegetation and the associated modification of the energy exchange) is considered a change of microclimate. When microclimatic changes are widespread there are usually associated local climatic changes.

There are some authors who maintain that world-wide changes in atmospheric circulation (that is, global climatic change) are causing displacement or extension of the boundaries of the desert regions (Lamb 1972). But there is no doubt that desertification is promoted and can be to some extent controlled by human activity, that is, by modification of the microclimate and other ground surface processes.

Erosion of soil is a major element. It follows from overgrazing, or overcutting of wood (often for fuel). The removal of vegetation by any agency has the effects outlined in the previous section. The loss of vegetation is usually only the end point of mismanagement in various respects. Lack of conservation practices, for example, insufficient control of gullying, failure to practice terracing or contouring during cultivation, improper drainage provisions, or lack of soil moisture retention procedures are all important.

Another element in desertification is associated, ironically, with water. Improper irrigation without provision of suitable drainage to allow for infiltration, results in water-logging (especially in little permeable soils). Evaporation then occurs and the consequent upward movement of water is followed by salt deposition at the surface. Such salinization is a major problem of arid and semi-arid climates, because of the infertility of soils so affected.

The passage of vehicles results in compaction of soil and a lowered permeability, making the soil more prone to water erosion during storms. Gully erosion is common too in association with roads where surface drainage water enlarges roadside ditches, in Mexico, for example, where annual precipitation is about 50 cm (Garduno in UNESCO, 1977). Desertification is an outstanding example of the interrelationship of the many aspects of processes at the earth's surface, involving the activities and concerns of man in high degree.

A fatalistic assumption that desertification is primarily due to events outside man's control, is unjustified. The consensus of expert opinion appears to be that, while climatic irregularity (for example successions of dry years, separating more favourable wet years) is important, this, in general, represents the 'normal' climate. Long term trends, effects over tens or hundreds of years (constituting climatic change in a more meaningful sense) are not easy to separate from the effects of contemporary, often more sudden, man-induced changes of the ground surface. The general effects of changing the properties of the ground surface, the modification of the surface by agriculture, timber cutting, or other human activities, have been well demonstrated. Rectification or at least the reduction of desertification is within the power of man, if these activities are restricted or carried out in a scientific manner. Two hurdles remain: the first concerns the necessity for education of those peoples inhabiting the areas prone to desertification. Frequently, there is not even the most elementary application of principles relating to soil erosion or moisture conservation, principles which relate to normal practice in drier regions of developed countries. The second concerns basic research. The depth of understanding of the microclimatic and other ground surface phenomena, and of the properties of earth materials under arid and semi-arid conditions, is too limited. The need for further research into the behaviour of earth materials in the temperate lands has been indicated repeatedly in this book; knowledge is particularly limited for those regions of the world which, because of their hostile or unstable conditions, have not been the location of sophisticated technologies. An analogy can be perceived with the cold or 'semi cold' regions. There the geotechnical challenges arising from the unstable ground conditions, relate not to the need for food and the practising of agriculture, but primarily to the need for energy for advanced industrial activity. Thus the problems relate to engineering, the construction of hydroelectric storage reservoirs and transmission lines, of gas and oil extraction facilities and pipelines, and similar structures.

The information most immediately required for these different types of 'marginal' lands is sometimes rather elementary, for example, the provision of numerical values of the more basic properties of earth materials. But ultimately the requirement is for a depth of understanding, similar to and including that produced only by decades of detailed study in the temperate regions, of the conditions and processes occurring at the surface of the earth.

Further reading

The books listed below, by chapter, are suggested because they provide further detail in fairly easily readable form (many of the works in the references are also valuable in this respect). The list is especially intended for the reader unfamiliar with one or more aspects of the diverse topics involved in studies of the earth's surface and its materials.

Chapter 1

Chorley, R. J., A. J. Dunn, and R. P. Beckinsale (1963, 1976) *The History of the Study of Landforms*, 2 Vols Methuen, 678 pp & 874 pp.

Davidson, D. (1978) *Science for Physical Geographers*. Arnold, 187 pp.

Flint, R. F. (1971) *Glacial and Quaternary Geology*. Wiley, 892 pp.

Gass, I. G., Peter J. Smith, and R. C. L. Wilson (eds) (1971) *Understanding the Earth*. Artemis Press (for Open University Press), 355 pp.

Holmes, Arthur (1965) *Principles of Physical Geology*. Nelson, 1288 pp.

Kellogg W. W. and R. Schware (1981) *Climate, Change and Society. Consequences of increasing Atmospheric Carbon Dioxide*. Westview, 178 pp.

Sears, F. W. and Mark W. Zemansky (1964) *University Physics* (3rd edn). Adison-Wesley, 184 pp.

Sparks, B. W. (1972) *Geomorphology* (2nd edn). Longman, 530 pp.

Wyllie, P. (1971) *The Dynamic Earth*. Wiley.

Chapter 2

Cox, K., (1971) *Minerals and rocks, in* **Gass, I. G., P. J. Smith, and R. C. L. Wilson (eds),** *Understanding the Earth*. Artemis Press, 355 pp.

Derbyshire E., K. J. Gregory and J. R. Haibs (1979) *Geomorphological Processes*. Dawson, Westview Press, 312 pp.

Embleton, C. and J. Thornes (eds) (1979) *Process in Geomorphology*. Arnold, 436 pp.

Firth, D. C. (1969) *Elementary Chemical Thermodynamics*. O.U.P. 136 pp.

Hunt, C. B. (1971) *Geology of Soils. Their Evolution, Classification and Uses*. Freeman, 344 pp.

Millot, George (1970) *Geology of Clays* (trans from French). Springer, 429 pp.

Strahler, A. N. and A. H. Strahler (1973) *Environmental Geoscience*. Hamilton Publ. Co. (Wiley), 511 pp.

West, R. G. (1968) *Pleistocene Geology and
Biology.* Longman, 377 pp.
White, R. E. (1979) *Introduction to the Principles
and Practice of Soil Science.* Blackwell, 198 pp.

Chapter 3

Barry, R. G. and R. J. Chorley (1976) *Atmosphere,
Weather and Climate.* Methuen, 432 pp.
Sellers, W. D. (1965) *Physical Climatology.*
University of Chicago Press, 272 pp.
Oke, T. R. (1978) *Boundary Layer Climates.*
Methuen, 374 pp.

Chapter 4

Bowles, J. E. (1979) *Physical and Geotechnical
Properties of Soils.* McGraw Hill, 478 pp.
Carson, M. A. and M. J. Kirkby (1972) *Hillslope
Form and Process.* C.U.P., 475 pp.
Craig, R. F. (1976) *Soil Mechanics.* Van Nostrand,
275 pp.
Lambe, T. W. and R. V. Whitman (1979) *Soil
Mechanics, S1 Version.* Wiley, 553 pp.
Pore Pressure and Suction in Soils (1961).
Butterworths, London.

Chapter 5

Clark, Sydney P. Jr, (1966) Thermal conductivity, *In
Handbook of Physical Constants.* Mem. 97, Geol.
Soc. Amer. 587 pp.
Jumikis, A. (1977) *Thermal Geotechnics.* Rutgers
U.P., 375 pp.
Wijk, W. R. van (ed.) (1963) *Physics of Plant
Environment.* North Holland Publ. Co., 2nd.
edn. 382 pp.

Chapter 6

Bolt, G. H. (1970) *Basic Elements of Soil Chemistry
and Physics. Pt. II: Soil Physics.* Post-Grad.
Training Program Soil Sci. Wageningen, 95 pp.
Hillel D. (1971) *Soil and Water.* Acad. Press. N.Y.,
288 pp.
Marshall, T. J. and J. W. Holmes (1979) *Soil
Physics.* C.U.P., 345 pp.
Nash, L. K. (1970) *Elements of Chemical

Thermodynamics.* (2nd edn.) Adison-Wesley, 184
pp.
Rose, C. W. (1966) *Agricultural Physics.* Pergamon,
230 pp.

Chapter 7

Edlefsen, N. E. and A. B. C. Anderson (1943)
Thermodynamics of soil moisture, *Hilgardia* **15**
(2), 298 pp.
Rose, C. W. (1966) *Agricultural Physics.* Pergamon,
230 pp.

Chapter 8

Chang, Jen-Hu (1973) *Climate and Agriculture.*
Aldine, 304 pp.
Glossary of Meteorology (1959) Huschke (Ed.) 638 pp.
Monteith, J. E. (ed.) (1975) *Vegetation and the
Atmosphere.* (2 vols) Academic Press, 278 pp.
(See esp. Lewis and Callaghar – Vol. 2).
Munn, R. E. (1966) *Descriptive Micrometeorology.*
Acad. Press. N.Y. 245 pp.
Oke, T. R. (1978) *Boundary Layer Climates.*
Methuen, 372 pp.
Rosenberg, N. J. (1979) *Microclimate: The
Biological Environment.* Wiley, 315 pp.

Chapter 9

Brunsden, D. (1979) Chapter 5; Mass movements, in
Embleton and Thornes (eds) *Process in
Geomorphology,* 436 pp.
Carson, M. A. (1971) *The Mechanics of Erosion.*
Pion, 174 pp.
Carson, M. A. and M. J. Kirkby (1972) *Hillslope
Form and Process.* C.U.P., 475 pp.
Coates, D. R. (ed.) (1977) Landslides, *Reviews in
Engg. Geol.* **III**, Geol. Soc. Amer., 278 pp.

Chapter 10

Chang, Jen-Hu (1958) *Ground Temperature.* Blue
Hill Observatory. Harvard University.
Rose, C. W. (1966) *Agricultural Physics.* Pergamon,
230 pp.
Wijk, W. R. van (ed.) (1963) *Physics of Plant
Environment.* (2nd edn.) North-Holland Publ.
Co., 382 pp.

Chapter 11

Andersland, O. B. and D. M. Anderson (eds.)
(1978) *Geotechnical Engineering for Cold Regions.*
McGraw-Hill, 566 pp.

Brown, R. J. E. (1970) *Permafrost in Canada; Its
Influence on Northern Development.* Univ.
Toronto Press, 234 pp.

Embleton, C. and C. A. M. King (1975) *Periglacial
Geomorphology.* Arnold, 203 pp.

Ferrians, O. J., R. Kachadoorian and G. W. Greene
(1969) Permafrost and related engineering
problems., *U.S. Geol. Surv. Prof. Paper 678,*
37 pp.

French, H. M. (1976) *The Periglacial Environment.*
Longman. 309 pp.

International Permafrost Conference, Proceedings,
1963; 1973; 1978. 1963 and 1973 Proceedings: Nat.
Acad. Sci., Washington. 1978 Proceedings: Nat.
Res. Coun., Canada.

Lachenbruch, A. H. (1968) Permafrost, in
Fairbridge, R. W. (ed.), *The Encyclopedia of
Geomorphology.* Reinhold, N.Y.

Washburn, A. L. (1979) *Geocryology. A Survey of
Periglacial Processes and Environments.* Arnold,
406 pp.

Chapter 12

Dregne, Harold E. (ed.) (1970) Arid Lands in
Transition. *Publ. 90 Amer. Ass. Adv. Sci.,*
Washington, 524 pp.

Greenland, D. J. and R. Lal (eds.) (1977) *Soil
Conservation and Management in the Humid
Tropics.* Wiley, 283 pp.

**Schwab, G. O., K. K. Barnes, R. R. Frevert and
T. W. Edminster** (1971) *Elementary Soil and
Water Engineering.* Wiley, 316 pp.

Yaron, B., E. Danfors and Y. Vaadia (1973) *Arid
Zone Irrigation.* Springer-Verlag, 434 pp.

References and bibliography

Aitchison, G. D., K. Russam and B. G. Richards (1967) *Engineering concepts of moisture equilibria and moisture changes in soils.* Road Res. Lab. Rept. 38 (Min. of Transp., UK), 36 pp.

Anderson, Duwayne M. and N. R. Morgenstern (1973) *Chemistry: Principles and Applications.* D.C. Heath, Lexington, Mass.

Anderson, Duwayne M., and N.R. Morgenstern (1973) *Physics, chemistry and mechanics of frozen ground: a review.* North Amer. Contrib. Permafrost 2nd Intern. Conf., Nat. Acad. Sci. Washington, 25–288.

Anderson, D. M. and A. R. Tice (1972) Predicting unfrozen water contents in frozen soils from surface area measurements (Frost Action in Soils), *Highw. Res. Rec.* 393, 12–18.

Baier, W. (1969) Concepts of soil moisture availability and their effect on soil moisture estimates from a meteorological budget, *Agric. Meteorol.* 6, 165–78.

Barry, R. G. and R. J. Chorley (1976) *Atmosphere, Weather and Climate* (3rd edn). Methuen, 432 pp.

Baver, L. D., W. H. Gardner and W. R. Gardner (1972) *Soil Physics* (4th edn). Wiley, 498 pp.

Beasley, R. P. (1972) *Erosion and Sediment Pollution Control.* Iowa State Univ. Press, 320 pp.

Beskow, G. (1935) *Tjälbildningen och tjällyftningen med särskild hänsyn till vägar och järnvägar.* Statens väginstitut, Stockholm. Meddelande, 48. (Also published as *Sveriges geologiska undersökning. Avhandlingar och uppsatser.* Serie C, 375, and trans. by Tech. Inst. North-Western Univ., Evanston, Ill., Nov., 1947.)

Birch, F. (1950) Flow of heat in the Front Range, Colorado, *Bull. Geol. Soc. Amer.* 61, 567–630.

Birkeland, P. (1974) *Pedology, Weathering and Geomorphological Research.* Oxford Univ. Press, 285 pp.

Bishop, A. W. (1960) *The Principle of Effective Stress.* Norw. Geotech. Inst. Pubn. 32, 1–5.

Bishop, A. W. and D. J. Henkel. (1969) *Measurement of Soil Properties in The Triaxial Test.* (2nd edn). Arnold.

Bjerrum, L. (1955) Stability of natural slopes in quick clay, *Geotechnique,* 5, 101–19.

Bjerrum, L. (1968) Progressive failure in slopes of overconsolidated plastic clay and shales, *Jour. Soil Mechs. Ground Divn., Amer. Soc. Civ. Eng.,* 93, SM5., 1–49.

Bjerrum, L., T. Loken, S. Heiberg and R. Foster (1969) A field study of factors responsible for quick clay slides, *Proc. 7th. Int. Conf. Soil Mech. Found. Engg.* 2 (Mexico), 531–40.

Bjerrum, L. and I. Th. Rosenquist (1956) Some experiments with artificially sedimented clays, *Geotechnique,* 6, 86–93.

Bolt, G. H. (1970) *Basic Elements of Soil Chemistry*

and Physics. Pt. II: *Soil Physics.* Post-Grad. Training Prog. Soil Sci., Wageningen, 95 pp.

Bowen, I. S. (1928) The ratio of heat losses by conduction and by evaporation from any water surface, *Phys. Rev.* **27**, 779–87.

Bozozuk, M. (1962) Soil shrinkage damages shallow foundations at Ottawa, Canada. *Engg. Jour.* **45** (7), 33–7.

Brady, N. C. (1974) *Nature and Properties of Soils* (8th edn). Macmillan, N.Y.

Brown, R. J. E. (1967) *Permafrost in Canada.* Map published by Div. Bldg. Res., Nat. Res. Coun. Can. (NRC 9769) and Geol. Surv. Can. (Map 1246A).

Brown, R. J. E. (1970) *Permafrost in Canada.* Univ. Toronto Press, 234 pp.

Brown, R. J. E. (1973) *Influence of climatic and terrain factors on ground temperatures at three locations in the permafrost region of Canada.* In Int. Permafrost Conf. 2nd Int. Permafrost Conf. 1973. Nat. Acad. Sci. Publn. 27–34.

Brown, W. G. (1964) Difficulties associated with predicting depth of freeze or thaw, *Can. Geotech. Jour.* **1** (4), 215–26.

Bryson, R. A. and T. J. Murray (1977) Climates of Hunger. Univ. of Wisconsin Press, 171 pp.

Bunting, B. T. (1967) *The Geography of Soil* (2nd edn). Hutchinson, 213 pp.

Burt, T. P. and P. J. Williams (1976) Hydraulic conductivity in frozen soils, *Earth Surface Processes*, **I** (4), 349–60.

Businger, J. A. and K. J. K. Buettner (1960) Thermal contact coefficient (A term proposed for use in heat transfer), *Jour. Meteorol.* **18**, 422.

Byers, R. B. (1965) *Cloud Physics.* Univ. Chicago Press, 191 pp.

Carson, M. A. (1971) *The Mechanics of Erosion.* Pion, London, 174 pp.

Carson, M. A. and M. J. Kirkby (1972) *Hillslope Form and Process.* Cambridge Univ. Press, 475 pp.

Carson, M. A. and D. J. Petley (1970) The existence of threshold hillslopes in the denudation of the landscape, *Trans. and Papers, Inst. Brit. Geogr.*, publin. 49, 71–95.

Cermak, V. (1971). Underground temperature and inferred climatic temperature of the past millenium, *Palaeogeography, Palaeoclimatology, Palaeoecology* **10**, 1–19.

Chandler, R. J. (1972) Periglacial mudslides in Vest-spitsbergen and their bearing on the origin of fossil 'solifluction' shears in low angled clay slopes, *Quart. Jour. Eng. Geol.* **5** (3), 223–41.

Chang, Jen-Hu (1968) *Climate and Agriculture.* Aldine, Chicago, 296 pp.

Chepil W. S. and N. P. Woodruff (1963) The physics of wind erosion and its control, *Adv. Agronom.* **15**, 211–302.

Coleman, J. D. (1949) Soil thermodynamics and road engineering, *Nature*, **163**, (4134), 143–5.

Cooke, R. U. and I. J. Smalley (1968) Salt weathering in Deserts, *Nature* **220**, 1226–7.

Cooke, R. U. and A. Warren (1973) *Geomorphology in Deserts.* Batsford, 374 pp.

Corte, A. E. (1966) Particle sorting by repeated freezing and thawing, *Biul. Perygl.* **15**, 175–240.

Craig, R. F. (1974) *Soil Mechanics.* Van Nostrand Reinhold, 275 pp.

Crawford, C. B. (1968) Quick clays of eastern Canada, *Eng. Geol.* **2** (4), 239–65.

Crawford, C. B. and G. H. Johnston (1971) Construction on permafrost, *Can. Geot. Jour.* **8** (2), 236–51.

Croney, D., J. D. Coleman and P. M. Bridge (1952) *The suction of moisture held in soil and other porous materials.* Road Res. Tech. Pap. 24 (DSIR, RRL, Harmondsworth, Middx.), 42 pp.

Curtis, C. D. (1976) Chemistry of rock weathering fundamental reactions and controls. In *Geomorphology and Climate* (Derbyshire, ed.). Wiley, London.

Czudek, T. and J. Demek (1970) Thermokarst and its influence on the development of lowland relief, *Quat. Res.* **1**, 103–20.

Davidson, D. A. (1978) *Science for Physical Geographers.* Arnold, 187 pp.

Davis, W. M. (1963) *Collected Essays* (reprinted). Dover.

Day, P. R., G. H. Bolt and D. M. Anderson (1967) Nature of soil water. In *Irrigation of Agricultural Lands.* Monogr. II, Agronom. Ser., Amer. Soc. Agronom., 1180 pp.

De Vries, D. A. (1952) *The Thermal Conductivity of Soil* (transl. from Dutch) Building Res. Stn, Libr. Comm., DSIR, No. 759, 1956.

Derbyshire, E. (1976) *Geomorphology and Climate* Wiley, 512 pp.

Eden, W. J. (1975) Mechanism of landslides in Leda Clay with special reference to the Ottawa area, *Proc. 4th Guelph Symp. Geomorph.*, pp. 159–71.

Edlefsen, N. E. and A. B. C. Anderson (1943) Thermodynamics of Soil moisture, Hilgardia, **15** (2), 298 pp.

Embleton, C. and J. Thornes (eds.) (1979) *Process in Geomorphology.* Arnold, 436 pp.

Everett, D. H. (1961) The thermodynamics of frost damage to porous solids, *Trans. Farad. Soc.* **57**, 1541–51.

FAO-UNESCO (1974) *Soil Map of the World.* Vol. 1, Legend. Paris, 59 pp.

Federer, C. A. (1975) Evapotranspiration, *Reviews of Geophysics and Space Physics*, **13** (3), 442–5.

Forsythe, W. E. (1969) *Smithsonian Physical Tables* (9th rev. edn). 827 pp.

Fredén, S. (1965) *Mechanism of frost heave and its*

relation to heat flow, 6th Int. Conf. Soil Mechs. Found. Engg., Vol. 1, 41–5.

French, H. M. (1974) Active thermokarst processes, Eastern Banks Island, Western Canadian Arctic, *Can. Jour. Earth Sci.* **11** (6), 785–94.

French, H. M. (1976) *The Periglacial Environment.* Longman, 309 pp.

Gates, V. M. (1965) Radiant Energy, its receipt and dispersal in Waggenar, P. (ed.) *Agricultural Meteorology*, Meteorol Monogr. 6, 28. Amer. Meteorol. Soc., Boston.

Gass I. G., Peter J. Smith and R. C. L. Wilson (eds) *Understanding the Earth.* Artemis Press, 355 pp.

Geiger, R. (1965) *The Climate Near the Ground.* Harvard Univ. Press, 611 pp.

Gidigasu, M. D. (1976) Laterite Soil Engineering *Developments Geotech. Engg.* **9** Elsevier, 544 pp.

Gillott, J. E. (1980) Use of the scanning electron microscope and Fourier methods in characterization of microfabric and texture of sediments, *Jour. Microsc.* **120** (3), 261–77.

Glantz, M. H. (ed.) (1977) *Desertification*, Westview, Boulder, Col. 346 pp.

Gold, L. W. and Arthur H. Lachenbruch (1973) *Thermal conditions in permafrost – a review of North American literature.* North American Contrib., 2nd Int. Permafrost Conf. 1973. Nat. Acad. Sci. Publn, pp. 3–25.

Goldring, R. (1971) Evolution in environments. In Gass, Smith and Wilson (eds) (1971), pp. 157–61.

Gradwell, M. W. (1968) The effect of grass cover on overnight heat losses from the soil, *New Zeal. Journ. Sci.*, Vol. II, 284–300.

Gribbin, J. (1976). *Forecasts, Famines and Freezes. Climate and Man's Future.* Wildwood House. 132 pp.

Griggs, D. T. (1936) The factor of fatigue in rock exfoliation, *Jour. Geol.* **44**, 781–96.

Grober, H., S. Erk, and U. Grigull (1961) *Fundamentals of Heat Transfer.* McGraw Hill, 527 pp.

Gupalo, A. I. (1972) *Thermal properties of the soil as a function of its moisture content and compactness.* NASA translation F-14364 (from *Pochvovedenie*, No. 4, 1959).

Gurr, C. G., T. J. Marshall and J. T. Hutton (1952) Movement of water in soil due to a temperature gradient, *Soil Sci.* **74**, 335–45.

Guymon, G. L. and J. N. Luthin (1974) A coupled heat and moisture transport model for Arctic soils, *Water Res. Res.*, **10** (5), 995–1001.

Gymer, R. G. (1973) *Chemistry: An Ecological Approach.* Harper and Row, N.Y., 801 pp.

Handbook of Chemistry and Physics (1976). Chem. Rubber Co. Press.

Hare, F. K. (1966) *The Restless Atmosphere* (4th edn). Hutchinson, 191 pp.

Harlan, R. L. (1974) Dynamics of water movement in permafrost: A review, *Proc. Workshop Seminar, Permafrost Hydrology*, Can. Nat. Comm., Int. Hydrol. Dec. pp. 69–77.

Heinemann, H. G. and R. F. Piest (1975) Soil erosion – sediment yield research in progress, *EOS Trans. Amer. Geophys. Union*, **56** (3), 149–59.

Hemming, C. E. (1965) Vegetation arcs in Somaliland, *Jour. Ecol.* **53**, 57–67.

Hillel, D. (1971) *Soil and Water. Physical principles and processes.* Acad. Press, N.Y., 288 pp.

Hjulstrom, F. (1935) Studies of the morphological activity of rivers as illustrated by the River Fyries, *Bull. Geol. Inst. Univ. Uppsala* **25**, 221–527.

Holmes, J. W. and J. S. Colville (1970) Forest hydrology in a Karstic region of Southern Australia, *Jour. Hydrol.* **10**, 59–74.

Horai, Ki-Iti (1971) Thermal conductivities of minerals. *Jour. Geophys. Res.*, **76**, 1278–308.

Horiguchi, K. and R. D. Miller (1979) Experimental studies with frozen soil in an 'ice sandwich' permeameter, *Cold Regions Science Technol.* **3**, 2–3, 177–183.

Idso, S. B., R. D. Jackson, R. J. Reginato, B. A. Kimball and F. S. Nakayama (1975) The dependence of bare soil albedo on soil water content, *Jour. App. Meteor.* **14**, 109–13.

ISSS (1963) Soil physics terminology. Commn. I (Soil Physics), *Bull. Int. Soc. Soil Sci.* **23**. 7–10.

ISSS (1975) Soil physics terminology. Commn. I (Soil Physics), *Bull. Int. Soc. Soil Sci.* **48**, 16–22.

Jackson, R. D. and D. Kirkham (1958) Method of measurement of the real thermal diffusivity of moist soil, *Soil Sci. Soc. Am. Proc.*, **22** (6), 479–82.

Jackson, Ray D. and Sterling A. Taylor (1965) Heat transfer, *Agronom. Monogr.* 9, Pt. 1. Amer. Soc. Agronom., 349–60.

Jackson, Ray D. and S. B. Idso (1975) Surface albedo and desertification, *Science* **189**, 1012–13.

Johansen, ø. (1972, 1973) *Beregningsmetode for varmeledningsevne av fuktige og frosne jordarter.* Del I, *Teoretisk grunnlag. Frost i Jord* **7**, 17–25. Del II, *Frost i Jord*, **10**, 13–28 (includes English summaries: *A method for calculation of thermal conductivity of soils*, Pt. 1 and 2).

Johansen, ø. (1975) *Varmeledningsevne av jordarter (Thermal conductivity of soils).* Inst. f. Kjøleteknikk, Trondheim. 231 pp. (with English summary, and includes large bibliography).

Johnson, J. P. (1973) Some problems in the study of rock glaciers, *Res. in Polar and Alp. Geomorph.*, 3rd Guelph Symp. on Geomorph., 84–149.

Journaux, A., J. P. Coutard *et al.* (1974) *Experiences de thermoclastie au Centre de Geomorphologie*, Bull. 18. Centre de Geomorph. de Caen, 31 pp.

Joynt, M. I. and P. J. Williams (1973) The role of ground heat in limiting frost penetration, *Symp. Frost Action in Roads*, Rept. Vol. 1, OECD, 189–203.

Judge, A. (1973) The prediction of permafrost thickness, *Can. Geotech. Jour.* **10** (1), 1–11.

Kaplar, C. W. (1965) Stone migration by freezing of soil, *Science* **149**, 1520–21.

Kaye, G. W. C. and T. H. Laby (1973) *Tables of Physical and Chemical Constants* (14th edn), Longman 386 pp.

Kenney, T. C. (1967) Sea level movements and the geologic history of the post-glacial marine soils at Boston, Nicolet, Ottawa and Oslo, *Geotechnique*, **14**, 203–30.

Keränen, J. (1929) Wärme- und Temperatur-verhaltnisse der obersten Bodenschichten. In *Einführung in die Geophysik*, II. Julius Springer, Berlin.

Kersten, M. S. (1949) Thermal properties of soils, *Engg. Exp. Stn*, Univ. of Minnesota, Bull. 28, 227 pp.

Kirkby, M. J. and M. A. Carson. (1972) *Hillslope Form and Process.* Cambridge Univ. Press, 475 pp.

Kojan, E. (1967) Mechanics and rates of natural soil creep. *Proc. 5th Ann. Engg. and Soils Engg. Symp.*, 233–53.

Koopmans, R. W. R. and R. D. Miller. (1966) Soil freezing and soil water characteristic curves, *Proc. Soil Sci. Soc. Amer.* **30** (6), 680–5.

Lachenbruch, A. H. (1962) Mechanics of thermal contraction cracks and ice-wedge polygons in permafrost, *Geol. Soc. Amer. Spec. Paper 70*, 69 pp.

Lachenbruch, A. H. (1963) Contraction theory of ice-wedge polygons: A qualitative discussion, *Proc. Int. Permafrost Conf.* Nat. Acad. Sci., Nat. Res. Council, Washington, 63–71.

Lachenbruch, A. H. (1968) Permafrost. In *The Encyclopedia of Geomorphology* (Fairbridge, R. W., ed.). Reinhold, N.Y.

Ladurie, E. L. R. (1972) *Times of Feast, Times of Famine. A History of Climate since the Year 1000.* Allen and Unwin, 428 pp.

Lamb, H. H. (1966) *The Changing Climate.* Methuen, 236 pp.

Lamb, H. H. (1972) *Climate: Present, Past and Future.* Methuen, 624 pp.

Lambe, T. W. (1958) Soil Testing for Engineers. Wiley, N.Y.

Lambe, T. W. and R. V. Whitman (1979) *Soil Mechanics* (SI version). 553 pp.

Lee, W. H. K. and S. P. Clark, Jr (1966) Heat flow and volcanic temperatures. In *Handbook of Physical Constants* (Clark, S. P., ed.). Geol. Soc. Amer., Mem. 97 (587 pp.), 483–511.

Leopold, L. B., M. G. Wolman and J. P. Miller (1964) *Fluvial Processes in Geomorphology.* W. H. Freeman, 522 pp.

Lerman, A. (1979) *Geochemical Processes Water and Sediment Environments.* Wiley, 481 pp.

Lockwood, J. G. (1978) The Climatic Future. *Progr. Phys. Geogr.* 2, 1, 107–115.

Löken, T. (1970) Recent research at the Norwegian Geotechnical Institute concerning the influence of chemical additions on quick clay, *Geol. For. Forhandl.* **92**, pt. 2, 133–47 (also as Norw. Geotech. Inst. Publn. 87, 1971).

Loughnan, F. C. (1969) *Chemical Weathering of the Silicate Minerals.* American Elsevier, 154 pp.

Lundqvist, Jan (1969) Earth and ice mounds: A terminological discussion. In *The Periglacial Environment* (Pewe, ed.) pp. 203–15. McGill-Queens Univ. Press., 487 pp.

Luthin, J. N. (ed.) (1957) *Drainage of Agricultural Lands.* Agronom. Monogr. 7, Amer. Soc. Agronom. Madison, Wisc., 620 pp.

MacFadyen, W. A. (1950) Vegetation patterns in the semi-desert plains of British Somaliland, *Geogr. Jour.* **116**, 199–211.

Mackay, J. R. (1970) Disturbances to the tundra and forest tundra environment of the Western Arctic, *Can. Geotech. Jour.* **7**, 420–32.

Mackay, J. R. (1973) The growth of pingoes, Western Arctic Coast, Canada. *Can. Jour. Earth Sci.* **10** (6), 979–1004.

Mackay, J. R. and R. F. Black (1973) Origin, composition and structure of perennially frozen ground ice: A review, *Proc. 2nd Int. Conf. Permafrost.* Nat. Acad. Sci., Washington, 185–92.

Marshall, C. E. (1977) *The Physical Chemistry and Mineralogy of Soils.* Vol II: *Soils in Place.* Wiley, 313 pp.

Mather, J. R. (1959) Determination of evapotranspiration by empirical methods, *Trans. Amer. Soc. Agric. Engineers* **2**, 35–8, 43.

Mather, J. R. (1974) *Climatology: Fundamentals and Applications.* McGraw Hill, 412 pp.

McRoberts, E. C. and N. R. Morgernstern (1974a) The stability of thawing slopes, *Can. Geotech. Jour.* **11** (4), 447–69.

McRoberts, E. C. and N. R. Morgernstern (1974b) Stability of slopes in frozen soil, Mackenzie Valley, N.W.T., *Can. Geotech. Jour.* **11**, 554–73. 554–73.

Miller, R. D. (1970) Ice Sandwich: Functional semi-permeable membrane, Science **169**, 584–5

Miller, R. D. (1975) Frost heaving in non-colloidal soils, *Proc. 3rd Int. Conf. Permafrost.* Nat. Res. Comm. Canada, I, 707–13.

Mitchell, J. K. and W. N. Houston (1969) Causes of clay sensitivity, *Jour. Soil Mech. Found. Div.*

95, SM3. Amer. Soc. Civ. Eng., 845–71.

Mitchell, J. K. (1976) *Fundamentals of Soil Behaviour*. Wiley, 422 pp.

Mitchell, R. J. and W. J. Eden (1972) Measured movements of clay slopes in the Ottawa area, *Can. Jour. Earth Sci.* **9** (8), 1001–13.

Moench, A. F. and D. D. Evans (1970) Thermal conductivity and diffusivity of soils using a cylindrical heat source, *Proc. Soil Sci. Soc. Am.* **34**, 377–81.

Monod, T. (1973) *Les Deserts*. Horizons de France, 247 pp.

Monteith, J. (1973) *Principles of Environmental Physics*. American Elsevier, N.Y., 241 pp.

Morgan, R. P. C. (1979), *Soil Erosion*. Longman, 113 pp.

Moum, J., T. Loken and J. K. Torrance (1971) A geochemical investigation of the sensitivity of a normally consolidated clay from Drammen, Norway, *Geotechnique* **21** (4), 329–40. (See also discussion: **22** (4), 675–6.)

Müller, Fritz (1959) Beobachtungen uber Pingoes. Detail-untersuchungen in Ostgrönland und in der Kanadischen Arktis, *Medd. om. Grönl.* **153** (3), 127. (Also as: *Observations on Pingoes*, Tech. Transl. 1073, Nat. Res. Counc. Can., 117.)

Munn, R. E. (1966) *Descriptive Micrometeorology*. Acad. Press, 245 pp.

Nakshabandi, G. A. and H. Kohnke (1965) Thermal conductivity and diffusivity of soils as related to moisture tension and other physical properties, *Agr. Met.* **2**, 271–9.

Nash, L. K. (1970) *Elements of Chemical Thermodynamics* (2nd edn). 184 pp.

Nersesova, Z. A. (1953–57) *In Materialy po laboratornym issledovaniiam merzlykh gruntov*. Inst. Merzlot., (Izd.) Akad. Nauk SSSR, Moskva, Sb. 1, 2, 3.

Oke, T. R. (1978) *Boundary Layer Climates*. Methuen, 372 pp.

Ollier, C. D. (1975) *Weathering*. Longman, 304 pp.

Paterson, W. S. B. (1969) *The Physics of Glaciers*. Pergamon, pp. 250.

Patterson D. and M. W. Smith 1981 The measurement of unfrozen water content by time domain reflectometry; results from laboratory tests. *Can Geotech. Jour.* 18, (1), 131–144.

Penner, S. S. (1968) *Thermodynamics for Scientists and Engineers*. Addison-Wesley, 288 pp.

Penrod, E. B., W. W. Walton and D. V. Terrell (1958) A method to describe soil temperature variation, *Jour. Soil. Mech. Found. Div., Proc. A.S.C.E.* 84.

Petrov, M. P. (1976) *Deserts of the World* (trans from Russian). John Wiley & Sons, 447 pp.

Price, L. W. (1973) Rates of mass-wasting in the Ruby Range, Yukon Territory, *Proc. 2nd Intern.*

Permafrost Conf. Nat. Acad. Sci., Washington, 235–45.

Priestley, C. H. B. (1959) *Turbulent Transfer in the Lower Atmosphere*. Univ. of Chicago Press.

Rapp, A. (1960) Recent development of mountain slopes in Karkevagge and surroundings, Northern Scandinavia, *Geogr. Annal.* XLII, 71–200.

Rapp, A., H. N. Le Houérou and B. Lundholm (eds) (1976) Can desert encroachment be stopped?, *Ecol. Bull.* 24. United Nations Envt. Progr. and Secret. for Intern. Ecology, 241 pp.

Rose, C. W. (1966) *Agricultural Physics*. Pergamon, 230 pp.

Rose, D. A. (1963) The effect of dissolved salts on water movement, *The Water Relations of Plants*, Symp. Brit. Ecol. Soc. Wiley, N.Y., 394 pp.

Rosenberg, N. J. (1965) The influence and implications of wind-breaks on agriculture in dry regions. In *Ground Level Climatology* (R. Sharp, ed.). Publn. 86, Amer. Assoc. Adv. Sci., 395 pp.

Rosenberg, N. J. (1974) *Microclimate: The Biological Environment*. Wiley, 315 pp.

Rosenquist, I. Th. (1966) Norwegian research into properties of quick clay – a review, *Engg Geol.* **1** (6), 445–50.

Rowell, D. L. and P. J. Dillon (1972) Migration and aggregation of Na and Ca clays by the freezing of dispersed and flocculated suspensions, *Jour. Soil Sci.* **23** (4), 442–7.

Royer, J. M. and G. Vachaud (1974) Determination directe de l'evapotranspiration et de l'infiltration par mesure des teneurs en eau et des succions, *Hydrol. Sciences Bull*, XIX, (3), (9) 319–36

Ryden, B. E. 1981 Hydrology of northern tundra. In *Tundra Ecosystems: A comparative analysis* (ed. L.C. Bliss *et al.*) C.U.P.

Sanchez, Petro A. (1976) *Properties and Management of Soils in the Tropics*. Wiley, 618 pp.

Sass, John H. (1971) The earth's heat and internal temperatures. In Gass, Smith and Wilson (eds) (1971).

Scheidegger, A. E. (1970) *Theoretical Geomorphology* (2nd edn.) Springer-Verlag, Berlin 435 pp.

Schofield, R. K. (1935) The pF of the water in the soil. Proc. *3rd Int. Cong. Soil Sci.* **2**, 37–48; **3**, 182–6.

Schofield, R. K. and J. V. Botello DaCosta (1938) The measurement of pF in soil by freezing point, *Agric. Sci.* **28**, 645–53.

Scott, R. F. (1964) Heat exchange at the ground surface, *Cold Regions Science and Engineering Monograph*. II-Al, US Army Cold Regions Research and Engineering Lab., Hanover, N. H.

Sears, F. W. and M. W. Zemansky (1965) *University Physics*. Addison-Wesley, 1028 pp.

Sellers, W. D. (1965) *Physical Climatology*. Univ. of Chicago Press, 272 pp.

Skaggs, R. W. and E. M. Smith (1968) Apparent thermal conductivity of soil as related to soil porosity, *Am. Soc. Agric. Engrs. Trans.,* **11** (4), 504–7.

Skaven-Haug, S. (1972) *The design of frost foundations, frost heat and soilheat*. Norw. Geotech. Inst. Publn. 90.

Skempton, A. W. (1953) Soil mechanics in relation to geology. *Proc. Yorks, Geol. Soc.* **29**, 33–62.

Skempton, A. W. (1960) Terzaghi's discovery of effective stress. In *From Theory to Practice in Soil Mechanics*. Wiley, N.Y., pp. 42–53.

Skempton, A. W. (1964) The long-term stability of clay slopes, *Geotechnique* **14**, 75–102.

Skempton, A. W. and F. A. Delory (1957) Stability of natural slopes in London Clay, *Proc. 4th Int. Conf. Soil Mech. Found. Eng.,* **2**, 378–81.

Skempton, A. W. and J. Hutchinson. (1969) Stability of natural slopes and embankment foundations, *Proc. Int. Soc. Soil Mech. Found. Eng.,* 291–339.

SMIC (Study of Man's Impact on the Climate) (1971) *Inadvertent Climate Modification*. M.I.T. Press, 308 pp.

Smith, M. W. (1975) Microclimatic influences on ground temperatures and permafrost distribution, Mackenzie Delta, Northwest Territories, *Can. Jour. Earth Sci.* **12** (8), 1421–38.

Smith, M. W. (1976) *Permafrost in the Mackenzie Delta, Northwest Territories*, Geol. Surv. Can, Paper 75–28, 34 pp.

Smith, M. W. and A. Tvede (1977) Computer Simulation of frost penetration beneath highways, *Can. Geot. Jour.* **14** (2), 167–79.

Somerton, W. H., J. A. Keese and S. L. Chu (1971) Thermal behaviour of unconsolidated oil sands, *Soc. Petr. Engrs. Journ.* **14** (5), 513–21.

Staple, W. J. (1967) Evaluation of flow parameters: Soil moisture; *Proc. Hydrol. Symp. 6.* Nat. Res. Counc. Canada, pp. 81–96.

Staple, W. J. (1969) Comparison of computed and measured soil moisture distribution following infiltration, *Soil. Sci. Soc. Amer. Proc.* 33, **6**, 840–7.

Statham, Ian (1977) *Earth Surface Sediment Transport*. Oxford, 184 pp.

Strakhov, N. M. (1967) *Principles of Lithogenesis*, Vol. 1 (Trans. J. P. Fitzsimmons, ed. S. I. Tomkeieff and J. E. Hemingway, Consultants Bureau, N.Y.). Oliver and Boyd, London.

Taylor, James A. (ed.) (1967) *Weather and Agriculture*. Pergamon, 225 pp.

Terzaghi, K. and O. K. Frohlich (1936) *Theorie der Setzung von Tonschichten*. Deutike, Leipzig. 166 pp.

Terzaghi, K. and R. B. Peck (1967) *Soil Mechanics in Engineering Practice*. Wiley. 729 pp.

Thornthwaite, C. W. (1948) An approach towards a rational classification of climate, *Geogr. Rev.* **38**, 55–94.

Thornthwaite, C. W. and F. K. Hare (1965) The loss of water to the air. *Met. Monogr.* **6** (28), 163–180.

Torrance, J. K. (1975) On the role of chemistry in the development and behaviour of the sensitive marine clays of Canada and Scandinavia, *Can. Geol. Jour.* **12**, 326–35.

Touloukhian, Y. S., R. W. Powell, C. Y. Ho and P. G. Klemens (1970) *Thermo-physical Properties of Matter*, Vol. 2: *Thermal Conductivity – Non-metallic Solids*. IFI/Plenum Data Corp., Washington.

Tricart, J. (1972) *The Landforms of the Humid Tropics, Forests and Savannas*. Longman, 306 p.

Tsytovich, N. A. (1975) *The Mechanics of Frozen Ground* (trans. from Russian). Scripta and McGraw Hill, 426 pp.

Turner, F. J. and J. Verhoogen (1960) *Igneous and Metamorphic Petrology* (2nd edn). McGraw Hill, 694 pp.

UNESCO (1977) *Desertification, its Causes and Consequences*. Pergamon, 448 pp.

U.S. Dept. Agric. (1960) *Soil Classification, a Comprehensive System (7th Approximation)*. Soil Conservation Service, 265 pp.

U.S. Dept. Agric. (1967) *Supplement to Soil Classification, a Comprehensive System (7th Approximation)*. Soil Conservation Service, 207 pp.

Vachuad, G., M. Vauclin and R. Haverkamp (1975) (1975) Towards a comprehensive simulation of transient water table flow problems. In *Modeling and Stimulation of Water Resources Systems* (G. C. Vansteenkiste, ed.). North Holland, pp. 103–118.

Van Duin, R. H. A. (1963) The influence of soil management on the temperature wave near the soil surface. *Institute of Land and Water Management Research, Tech. Bull.* No. 29, 21 pp.

Vialov, S. S. (ed.) (1965) *The strength and creep of frozen soils and calculations for ice-soil retaining structures*. Translation 76. US army, Cold Reg. Res. Engg. Lab. Hanover, N.H., 301 pp.

Von Arx, W. S. (1974) Energy: Natural limits and abundances, *EOS, Trans. Amer. Geophys. U.* **55** (9), 828–32.

Wahrhaftig, C. and A. Cox (1959) Rock glaciers in the Alaska range, *Bull. Geol. Soc. Amer.* **70**, 383–436.

Ward, R. C. (1975) *Principles of Hydrology*. McGraw Hill, 367 pp.

Warren, A. (1979) Aeolian processes. In *Process in Geomorphology* (Embleton C. and J. Thornes) 436 pp.

Washburn, A. L. (1956) Classification of patterned ground and review of suggested origins, *Bull. Geol. Soc. Amer.*, **67**, 823–66.

Washburn, A. L. (1967) Instrumental observations of mass wasting in the Mesters Vig district, Northeast Greenland, *Medd. om Gronl.* **160** (4), 318 pp.

Washburn, A. L. (1979) *Geocryology. A Survey of Periglacial Processes and Environments*. Arnold, 406 pp.

Weller, G. and B. Holmgren (1974) The Microclimates of the Arctic Tundra, *Jour. Appl. Meteor.* **13**, 854–62.

Wijk, W. R. van (ed.) (1966) *Physics of Plant Environment* (2nd edn). North-Holland, 382 pp.

Williams, G. P. (1970) The thermal regime of a sphagnum peat bog, *Proc. 3rd Int. Peat Congr.*, 195–200.

Williams, P. J. (1957) Some investigations into solifluction features in Norway, *Geogr. Jour.* **123** (2), 42–58.

Williams, P. J. (1959) The development and significance of stony earth circles. *Skr. d. Norske Vidensk. -Akad., I. Mat. Naturv. Kl.*, No. 3, 3–14.

Williams, P. J. (1961) Climatic factors controlling the distribution of certain frozen ground phenomena, *Geogr. Ann.* Vol. LXIII (3–4), 339–47.

Williams, P. J. (1963) Canadian investigations on soil erosion. *Nach. d. Akad. Wiss. i. Gött, II Math. -Phys. Kl,* Nr. 18, 275–77.

Williams, P. J. (1966) Downslope soil movement at a sub-arctic location with regard to variation with depth, *Can. Geot. Jour.* **3** (4), 191–203.

Williams, P. J. (1967) *Properties and Behaviour of Freezing Soils*. Norw. Geotech. Inst. Publ. 72, 120 pp.

Williams, P. J. (1967) Replacement of water by air in soil pores, *The Engineer*, 223 (5796). 293–8.

Williams, P. J. (1968) Ice distribution in permafrost profiles. *Can. Jour. Earth Sci.* **5** (12), 1381–6.

Williams, P. J. (1975) *Report on investigation of upward heat flows to the frost line*. Ont. Min. Transp. Comm., 22 pp.

Williams, P. J. (1976) Isothermal volume change in frozen soils. *Laurits Bjerrum Mem. Vol.*, Norw. Geot. Inst., Oslo, 233–46.

Williams, P. J. (1979) *Pipelines and Permafrost*. Longman, 108 pp.

Williams, P. J. and W. G. Nickling (1972) Ground thermal regime in cold regions, *Proc. 2nd Guelph Symp. Geomorph.*, 27–44.

Wilson, C. V. (1975) *The climate of Quebec: energy considerations Climatological Studies* 23. Environment Canada, 120 pp.

Withers, Bruce and Stanley Vipond (1980) *Irrigation Design and Practice* (2nd edn). Cornell University Press, 300 pp.

Wood, J. A. (1979) Effect of selected physical parameters on the thermal conductivity of soils. Unpublished B.Sc. thesis, Carleton University.

Woodside, W. and J. H. Messmer (1961) Thermal conductivity of porous media I. Unconsolidated sand, *J. Appl. Phys.* **32** (9), 1688–1706.

Wyllie, P. J. (1976) *The Way the Earth Works. An Introduction to the New Global Geology and its Revolutionary Development*. Wiley, 296 pp.

Yair, A. and H. Lavee (1976) Run-off generative process and run-off yield from arid talus mantled slopes, *Earth Surf. Proc.* **1** (3), 235–48.

Yaron, B., E. Danfors and Y. Vacidial (eds) (1973) *Arid Zone Irrigation*. Springer, 434 pp.

Index